The International Library of Bioethics

Founding Editors

David N. Weisstub

Thomasine Kimbrough Kushner

Volume 100

The *International Library of Bioethics* – formerly known as the International Library of Ethics, Law and the New Medicine comprises volumes with an international and interdisciplinary focus on foundational and applied issues in bioethics. With this renewal of a successful series we aim to meet the challenge of our time: how to direct biotechnology to human and other living things' ends, how to deal with changed values in the areas of religion, society, and culture, and how to formulate a new way of thinking, a new bioethics.

The *International Library of Bioethics* focuses on the role of bioethics against the background of increasing globalization and interdependency of the world's cultures and governments, with mutual influencing occurring throughout the world in all fields. The series will continue to focus on perennial issues of aging, mental health, preventive medicine, medical research issues, end of life, biolaw, and other areas of bioethics, whilst expanding into other current and future topics.

We welcome book proposals representing the broad interest of this series' interdisciplinary and international focus. We especially encourage proposals addressing aspects of changes in biological and medical research and clinical health care, health policy, medical and biotechnology, and other applied ethical areas involving living things, with an emphasis on those interventions and alterations that force us to re-examine foundational issues.

Emma Tumilty · Michele Battle-Fisher
Editors

Transhumanism: Entering an Era of Bodyhacking and Radical Human Modification

 Springer

Editors
Emma Tumilty
Institute for Bioethics and Health
Humanities
University of Texas Medical Branch
Galveston, TX, USA

Michele Battle-Fisher
Boonshoft School of Medicine
Wright State University
Dayton, OH, USA

ISSN 2662-9186   ISSN 2662-9194 (electronic)
The International Library of Bioethics
ISBN 978-3-031-14327-4   ISBN 978-3-031-14328-1 (eBook)
https://doi.org/10.1007/978-3-031-14328-1

This Springer imprint is published by the registered company Springer Nature Switzerland AG
The registered company address is: Gewerbestrasse 11, 6330 Cham, Switzerland

Contents

Editors and Contributors

About the Editors

Dr. Emma Tumilty is feminist bioethicist and Assistant Professor& Associate Director of Graduate Studies in the Institute for Bioethics and Health Humanities at the University of Texas Medical Branch. Her interests include research ethics, translational medicine, and innovation, as well as health justice and access. She sees biohacking and related activities such as open medicine, SynBio, biopunk practices, etc. as offering exciting opportunities for discussion of how science and medicine are done and by who. Her interest in transhumanism is focused on those seeking to disrupt the status quo with human+technology exploration—the crip technoscience writers, feminist technomaterialists, afrofuturists, etc. who imagine different communities, morphological freedoms, and futures facilitated by technological integration. She is an Associate Editor for the Journal of Empirical Research in Human Subjects Research and the journal: Progress in Community Health Partnerships, as well as the Book Review Editor for the International Journal of Feminist Approaches to Bioethics. She has published widely in research ethics, bioethics generally, and health service research.

Michele Battle-Fisher is an Adjunct Assistant Professor at the Wright State University Boonshoft School of Medicine and an Adjunct Instructor at Temple University Lewis Katz School of Medicine. She is the author of *Application of Systems Thinking to Health Policy and Public Health Ethics: Public Health and Private Illness* (Springer), a 2016 Doody's Core Title selection. She is a systems science & public health researcher as well as bioethicist. Her scholarship ranges from public health/health disparities to systems science/complexity theory and their application to health. Her academic scholarship has been published in systems science, bioethics, humanities and communication studies peer-reviewed journals and academic outlets. She is a member of The Bertalanffy Center for the Study of Systems Science (BCSSS). She is also a Research Scholar at the Ronin Institute. She was a speaker at TEDxDartmouth 2018 where she discussed the "Paradigm Shift" of the Health

Systems Science curriculum in health and clinical medicine. She was a Visiting Scholar at the Hastings Center.

Contributors

Auer Robin Markus Institut für Anglistik und Amerikanistik, Braunschweig, Germany

Earle Joshua Virginia Polytechnic Institute and State University (Virginia Tech), Blacksburg, USA

Ivic Sanja Institute for European Studies, Belgrade, Serbia

Karakasis Georgios University of the Basque Country, Donostia-San Sebastian, Spain

Kendal Evie Swinburne University of Technology, Melbourne, VIC, Australia

Marshall Belinda University of St. Andrews, St. Andrews, UK

McFarlane Steven Madison, United States

Reverter Sonia Castelló de la Plana, Spain

Stendera Marilyn University of Texas Medical Branch, Galveston, US

Tumilty Emma Institute of Bioethics and Health Humanities, University of Texas Medical Branch, Galveston, USA

Wilson Aaron Bruce South Texas College, McAllen, US

Chapter 1
Introduction

Emma Tumilty

> *Meanwhile, in the real world, as a direct consequence of increasingly audacious moves by some scientists to engineer future generations, important decisions must now be made— decisions that will set a new course for science, society, and humanity. May these decisions be inclusive and consensual. May they be characterized by wisdom and benevolence. And, may we never lose sight of our responsibilities to us all.*
>
> *(Francois Baylis, Altered Inheritance, p. 221)*

Jillian Weise, a self-proclaimed cyborg (they have a computerised prosthetic limb), writer, and disability activist, calls all non-disabled people trying to be cyborgs, tryborgs (Weise 2016). For Weise disabled people are already cyborg and experience living with embedded and embodied technology all the time. This is especially the case if included in what we count, as technologies, are pharmaceutical interventions, as the first people thinking of cybernetic organisms imagined (Williams 2019). Those now using technologies as accessories, who do not depend on them for their existence or key parts of their daily lives, but just dabble and explore enhancements to be futuristic are, according to Weise, fakers. These fakers promoting transhumanism, biohacking and cyborg futures rarely recognise the real life cyborgs all around them. Weise (2016) and others (Williams 2019) correctly note, that seldom are the already present cyborgs amongst us invited into the discussion; they are rarely included on the expert panels on transhumanism and cyborg futures despite their obvious expertise. Morphological freedom…but no, not like that.

The media often portrays transhumanism and biohacking as the fantasies of rugged (and rogue) libertarians pursuing bigger, better, faster, smarter, stronger humans. Their dreams seem hyper-individualistic, narrowly performance-focused in a disappointingly neoliberal way or alternatively are desires for everlasting youth and immortality (mostly of the rich) and leave a kind of bitter aftertaste. It is no

E. Tumilty (✉)
Institute of Bioethics and Health Humanities, University of Texas Medical Branch, Galveston, USA
e-mail: emtumilt@utmb.edu

wonder then that writers like Melinda Hall describe transhumanism (and enhancement activities such as biohacking) as inherently eugenic (Hall 2016). For Hall,
debating the blurred lines (if present at all) between enhancement and treatment is
arbitrary; recognising that all transhumanism rejects fundamental elements of being
human such as vulnerability, dependency and community is key (Hall 2016).

To read Hall and reject technological enhancement of human bodies and lives fails
to recognise as Weise points out that disabled people are actually at the forefront of
human + technology hybrids. Hall's accurate analysis of the ableist imaginings of
the dominant voices in transhumanism calls for urgent intervention not wholesale
dismissal. As Williams (2019) explains, transhumanists dreaming of technologically
enhanced humans for utopian futures and space travel would do well to engage with
and learn from disabled people.

Ableism and eugenics though bad enough are not the only objections to biohacked
and transhumanist futures. Bioconservatives foresee a range of dystopias resulting
from transhumanist activities. Objections range from those occupied with the nature
of being human and what might be lost with greater technological integration,
enhancement, hybridity, fearing some form of dehumanisation (Bostrom 2011);
to those worried about worlds in which transhumanism creates further class divisions that may be insurmountable due to who can and cannot access technological
developments (Porter 2017).

Others imagine that transhumanism and biohacking offer us opportunities and
tools to be disruptive to transform the dominant exploitative and discriminatory
scripts of society and create new ways of being together (Laboria Cuboniks 2018).
They argue that biopunk and biohacking practices not only democratise science and
medicine in ways that improve access, but allow for a greater diversity of ideas
to come to fruition (Wohlson 2012; Laboria Cubonicks 2018) and can unsettle the
harmful and hegemonic relations of white supremacy, capitalism, and patriarchy.
They argue that current practices, processes, and priorities of establishment science
and medicine can only ever be limited and limiting because of the hierarchical and
discriminatory nature of research and development both in the public and private
sphere with funding allocated to priorities identified by the few (Wohlson 2012).
Democratising who has what knowledge and what they can do with it, not only has
implications for the individual, but can shift power, build capacity within communities and societies, and disturb the dominance of current ideas and gatekeeping of what
constitutes living a good life: flourishing. In fact, some have theorised that biohacking
and science activities modelled on open source software communities can provide a
different kind of justice, generative justice (Eglash 2016; Eglash et al. 2017). Eglash
argues that biohacking communities built on open source philosophies provide space
for greater creativity and wellbeing. Such communities would generate the things
they want or need and directly access and enjoy them without an exploitative, discriminatory, and/or profit-driven middle-step (2016); not only that but if created successfully these communities could overcome the "cyclical social damage" resulting from
the underrepresentation of marginalised folks in STEM (2017). Such communities
build and share capacity, knowledge, and resources. Therefore, while some argue
transhumanism is ableist and possibly oppressive or marginalising, biohacking as a

set of a practices used by diverse communities could have the potential to generate new relations between people and within communities and society. Some of this we do not have to imagine, disabled led maker-spaces creating open-source prosthetics, disability-aids, etc. already exist and thrive (Bosse and Pelka 2020; Bijadi et al. 2017; as just some examples) and Xenofeminists talk of communities sharing practices and technologies for gender disruption amongst other things (Laboria Cuboniks 2018).

The authors in this book touch on many of these arguments of both bioconservatist worries and the hopes of disrupting the status quo; sometimes daring to dream of something better, while others advise caution.

Some terminology for new readers to this kind of work. Transhumanism has a 'big tent' kind of problem. While in general, transhumanism can be defined as a philosophy that seeks to transcend the natural state of being human with technology to improve performance, avoid suffering and cheat death, the philosophers within this tent are performing a variety of acts. From libertarians to socialists, entrepreneurs to anarchists, feminists to autocrats (More and Vita-More 2013; Manzocco 2019). Transhumanists can espouse a range of philosophical and political positions that affect their conception of transhumanism and transhuman/posthuman futures. Whether transhumanism is seen as a way to improve equity and social justice or the ability for individuals to pursue personal goals, it all, at its core includes the connection of the human with the technological in ways that are thought to deeply change what it means to be human.

Biohacking is another concept and set of practices that covers a lot of ground. While there are now endless podcasts, and YouTube videos of "biohacks" from Silicon Valley regarding bullet coffee, breathing, cold showers and how to work with your vagus nerve for optimum performance (Lee 2015; Grewe-Salfield 2021), biohacking early on encompassed more radical and outsider practices (Duarte 2014). These overlap with other activities such as bodyhacking and grinding and crossover with other subcultures. These more radical biohacks involved imaginative and riskier interventions. Whether it was implanting magnets in fingertips for haptic responses or assistive devices, or RFIDs, and biometric devices[1] or using eye drops for night vision (Dvorsky 2015), or more recently trying to edit your own genes (Lee 2017).

Others in the transhumanism space believe that immortality comes not through genetic engineering and human-technology hybrids, cyborgs, etc. but through the connection of human awareness with artificial intelligence—our brains and computers becoming one (Edman 2019). On the path to this, as various technology develops, others imagine situations in which technologically enhanced humans, cyborgs, robots and androids all live together, representing various stages of technological advancement on the pathway to the singularity (Barfield 2015). What might this world look like, how would we relate to each other, would we relate to each other—is everyone in this equation an "Other?" This raises questions of what it means to be human; whether being human has some special value that needs to be

---

[1] Various options offered by Grindhouse Wetware and on show at bio/body-hacking conferences in attendees.

accounted for in how we create and relate; and what things we can and should pursue as editable/assist-able/enhance-able in our human bodies, minds, and experiences.

In the past transhumanism was a largely a theoretical discussion occupied with utopian/dystopian projection and science fiction (Edman 2019). The twenty-first century has seen greater progress in robotics, brain-computer interfaces, artificial intelligence, as well as the spread of more community DIY Bio labs and maker spaces, as well as biopunks attempting to make more and more tools accessible (Wohlson 2012). While Fukuyama famously called transhumanism the most dangerous idea for humanity (Fukuyama 2004), he's not the only one to sound alarm bells about what our tinkering with ourselves might lead to, whether that is related to genetic modification/engineering (Baylis 2019) or connecting our brains to the web (Trimper et al. 2014). Endless possibilities bring endless potential threats and harms. To think through where we might be going, a better understanding of what transhumanism is and what the possible implications of transhumanism and biohacking are is needed.

This book brings together authors trying to understand transhumanism and those imagining what we might need to do if we want a biohacked and transhuman (and eventually posthuman) world to be better for everyone. Earlier authors raise bioconservative worries about the nature of being human while later authors worry about what transhuman realities may look like and what they could look like. The authors think about the issues through a variety of foci using different disciplinary knowledges and skills but there is a constant thread of democratisation of the tools for self- and world-building, relational aspects of being, ideas of disrupting current hierarchies, and discussion of vulnerability and care.

Our first author, Aaron Wilson (Chap. 2), puts forward the idea that pragmatism has much to offer in understanding transhumanism as a philosophy and philosophical approach. Using central pragmatist thinkers such as Pierce, James, and Rorty, Wilson explains pragmatist theory as naturalistic, evolutionistic, and melioristic in its view of humans and society, supporting improvement of the human condition through scientific advancement and democracy. Wilson argues that pragmatism through its parallels with transhumanist thought offers transhumanism support for its prescriptive claims by rationalising the pursuit of its goals. He offers here the why of transhumanism, rather than the what and how.

Stendra in Chap. 3, puts forward enactivist philosophy as a way to both understand and problematize understandings of transhumanism. Enactivism understands cognitive agents to be autonomous embodied beings in constant interaction with their surroundings in ways that are both positive and negative. For enactivists there is a delicate balance between a system constantly adapting, developing, and changing itself and change so stark that it fundamentally undermines what the original agent was, leading to disintegration (and death). In enactivism, the body is central to a cognisers' being. Enactivism serves therefore as a helpful tool in exploring transhumanism. What might we understand better if we apply integration/disintegration in enactivism to transcendence of humanity? Stendra focuses on organisational integrity, embodiment, and precarity, to go back and forth between enactivist theory and tranhumanism to stimulate our thinking about what morphological freedom means for our conceptual understandings of ourselves and our relationships with each other, and the world around us.

Thinking through what about our nature is essential, Karakasis in Chap. 4 presents what might be a considered a more bleak view of transhumanism. He investigates transhumanism through Heideggerian thought. The merging of the organic and technological, to overcome decay and death pushes the human into a new state of being and this new state is the ultimate embodiment of the Heideggerian 'will to life.' This chapter calls into question what role death and decay play in the definition of human and human being. How does our vulnerability and mortality define us as human? If relationships and community are central to what defines us as human, then what might these look like for humans who are immortal. Karakasis much like Hall (discussed earlier in this Chapter) drawing on Heidegger believes that death serves as an important anchor or reference point for existence and experience and without it, we may lose an important component needed to generate meaning in our lives. Therefore, in transcending the human through the elimination of death, do we lose something vital to meaning-making?

Adding further to a discussion of what might be necessary to our nature and functioning, Chap. 5 moves from the individual to the collective. Robin Auer walks us through what the implications of this new kind of embodiment might actually be. Drawing on what we already know about the use of technological tools by humans, in relation to how it can support and change culture, divide communities, etc. Auer asks the hard questions in relation to how transhumanism may affect the social fabric of our societies. Going beyond the discussion of access and inequality and how this could cause fractured societies, Auer persuasively argues, that even with equal access, the diversity of morphological freedom autonomously chosen may break down our ability to have shared experiences. Shared experiences that may be necessary for society to be a society. Auer argues that without consideration of these questions unbridled transhumanism could threaten society itself not just transcend the human. Can transhumans exist in a post-societal world? These ideas are crucial to how we think about different beings and their rights and how we relate and interact. As discussed earlier it is likely we will live in a world where various agents (cyborgs, androids, artificially intelligent robots, etc.) exist at once, that is, we will not all become transhuman or connect our minds to artificial intelligence at the same time (Barfield 2015). Thinking through what that means for our institutions, our relationships, our everyday is important.

In the chapters so far, what lurks just beyond the periphery of the arguments, hiding in the shadows, seems to be the assumed irreversibility of biohacking, enhancement, and transhumanism. Once hacked, there is no going back; once we pursue the transhumanist goal, our course is set. Marshall in Chap. 6 tries to reverse our course. After making arguments about what we might lose and what problems can arise from embedding technology in ways that affect our ongoing embodiment and interaction with the world around us, Marshall suggests rather than creating enmeshed human-technology hybrids, why not fulfil some of transhumanism's goals through virtual reality without the irreversiable commitments. Virtual reality can offer many of the experiences we may want through enhanced bodies but in ways that we can simply stop engaging in and take off. Boundaries between work and play that could be blurred with the bodyhacking discussed in Kendal's chapter on work (Chap. 7), could be better maintained using virtual and augmented realities rather than integrated technology in our bodies. The chapter uses a variety of literature to make an argument for virtual reality avatars instead of cyborg or biohacked bodies. It has particular relevance as we begin to discuss the Metaverse and Web 3.0. Others have discussed how avatars in virtual space provide humans with experiences that can be thought of as "real" even when having multiple avatars doing things that one might be unable to do in reality (Bainbridge 2013). One person can be a collection of agents in virtual space. Whether transhumanists focused on life extension and never-ending youth will find this persuasive is questionable, but that does not mean that others will not.

Marshall's justifications for thinking of virtual reality call in part on our inability to switch off from work. Corporate capture of transhumanist technologies and development could lead to very dystopic societies, and as Kendal explores in Chap. 7, workplaces. It is conceivable to think of workplace enhancement arms races; competitive settings promoting use of biohacks to get ahead. Using a bioethical analysis, Kendal discusses the possibilities of enhanced humans in the workplace and potential pitfalls, including careful analysis of issues such as autonomy, regulation and the argued inevitably of these futures. Kendal also raises the possibility of enhancements helping us to work less, potentially transforming our relationship to work and leisure. Xenofeminists also posit that biohacking can be disruptive of our personal and societal relationships to work and labour (Hester 2018). At the very least biohacking and maker-spaces could be places where people and communities devise countermeasures to workplace and societal surveillance and exploitation through technological solutions (and in some ways already are) (Ming 2020).

What is it we want is really the core question of many of these problems. How will our merging with technology affect our autonomy, our agency? How might we want it to affect these things? In Chap. 8, McFarlane explores enhancement targeted at agency itself. By explaining different types of agency and the way they affect our lives, McFarlane walks us through the possibilities of enhancing agency and their implications. He speculates that unlike other enhancements, something that could help one stay on track with one's goals is likely to be much more widely popular. If you could stick to your exercise regime, hold your boundaries, or make sure you write that paper, by taking a pill, or implanting a chip, sitting ten minutes a day with a brain-device on of some kind (all speculation, no agential enhancements yet exist), wouldn't you? How could this be detrimental? McFarlane raises possible objections and worries to such interventions and much like Kendal in Chap. 7 ends with some hopefulness about what the future might hold in this space.

In Chap. 9, Joshua Earle further explores this constant "two sides of the same coin" problem of transhumanism and biohacking where the same aspect can be imagined with both positive potential and negative outcomes in relation to morphological freedom. Morphological freedom is one of the central tenets and rights described by transhumanists and cyborgs.[2] It states that a person has the right to create themselves (or not) as they see fit. The idea being that as a society we should accept all comers, no matter their appearance, capabilities, technological enhancements, etc. Earle discusses what morphological freedom seems to mean in today's world and what this foreshadows for the future of transhumanism without more serious thought given to the importance of relationships, community, and care. Earle describes three different communities that have morphological differences, their relationships to each other, and how they are treated by the world around them. Earle makes clear the impoverished thinking of hyper-individualistic morphological freedom, but positively puts forward conditions that could make a biohacked future one we might actually want.

The question of what consideration different beings are due based on the nature of their being is a fundamental one to a morphologically free future. In Chap. 10, Ivic argues that postmodernist thought about subjectivity is the only path forward for transhuman and posthuman ethics. Ideas of subjectivity within postmodern thinking, because of their nature, offer the only solution to transhumanists who need an ethical acceptance of (and focus on) Otherness. If there is hope of disrupting what seems to be so very human—binary hierarchies and harmful power relations—then it is through alternative understandings of subjectivity and acceptance of the resulting implications.

Further imagining new power relations, Reverter in Chap. 11, wonders whether transhumanist practices may unsettle gender binaries and sexual identities in a way that could make a transhumanist world inherently non-sexist. Reverter argues that dominant discourses in transhumanist spaces seem to reinforce rather than disrupt oppressive gender scripts. Drawing on Haraway, Bradiotti and Xenofeminism, she explores the potential for transhumanism to create a non-binary world. Quite practically, Reverter makes clear what our democratic public institutions need to be doing in response to transhuman and biohacking activities now, in order to ensure that the futures we create are not technologically enhanced versions of our current oppressive relations, but help us to create equality and freedom.

One of the key ways that gender and sex are frequently discussed both in transhumanism, and society in general, is in relation to reproduction. In our final formal chapter (more on the secret reading in this book below), Kendal provides an extensive and comprehensive discussion of ectogenesis and other emerging reproductive technologies in relation to transhumanism, but especially in the context of space travel. Chapter 12 offers an overview of the range of arguments for and against a variety of practices and what their implications are, not only our future, but even our concepts of human and family. Again, using a bioethical analysis, she outlines the different ways these practices are thought to be justifiable (or not) in different contexts, with space exploration presenting greater ethical justification for practices that at present

---

[2] A couple of examples of this: https://www.cyborgfoundation.com/ created by Neil Harbisson a cyborg artist with achromatopsia who has an implant that translates colour into sound so that he can experience colour in this world. Or the traditional Transhumanism Manifesto by Nathasha Vita-Moore (now version 4) https://natashavita-more.com/transhumanist-manifesto/.

are rejected. Space is a hostile environment for humans, if we do not change what we are then our efforts to explore the next frontier are likely to be futile.

Like any 90s kid will know, every good (CD) album has a hidden track. In the back matter of this book, taking the focus of space exploration further, Chris Hables Gray a key thinker in transhumanism and discussions of the singularity, has written a short story about the experience of space with augmentation.

Throughout this book, the various authors illuminate aspects of transhumanism and biohacking in ways that expand the conversation about how we should be thinking about our possible futures. One of the key features of our humanity alongside those things discussed here such as relationships, vulnerability, and care has been tool use. Are integrated tools, melding biology and technology that radically different, or just a natural progression of our innate nature to improve our situation? Whether we will ever "upload" our minds to the web to merge with artificial intelligence remains to be seen, but it does seem much more possible in the twenty-first century than it did even last century. What is clear, is that implants, enhancements, augmentations, pharmaceuticals and a range of other hacks are already here and already being used. Genetic-engineering and modification appears to be steaming forward. The future will be decided by who we let take the wheel. If that is the only the libertarian entrepeneurs of the world then we will get more of the same, just at a more intense pace—the same exploitation, oppression, and extraction of human potential for mundane profit-making that benefits a few. If we can create multiple vehicles, multiple wheels on multiple tracks, through community building and capacity building, then we might at least have a variety of futures we can engage with…

"If God is Change, then... then who loves us? Who cares about us? Who cares for us?"

"We care for one another," I said. "We care for ourselves and one another."

(Octavia Butler, Parable of the Talents, p. 378)

References

Bainbridge, W. 2013. Transavatars. In *The transhumanist reader: Classical and contemporary essays on the science, technology, and philosophy of the human future*, ed. Max More and Natasha Vita-More. Wiley.

Barfield, Woodrow. 2015. *Cyber-humans: Our future with machines*. Springer.

Baylis, Françoise. 2019. *Altered inheritance*. Cambridge, Massachusetts: Harvard University Press.

Bijadi, Sachin, Erik de Bruijn, Erik Y. Tempelman, and Jos Oberdorf. 2017. Application of multimaterial 3D printing for improved functionality and modularity of open source low-cost prosthetics: A case study. In *Frontiers in biomedical devices*, vol. 40672, p. V001T10A003. American Society of Mechanical Engineers.

Bosse, Ingo Karl, and Bastian Pelka. 2020. Peer production by persons with disabilities–opening 3D-printing aids to everybody in an inclusive MakerSpace. *Journal of Enabling Technologies*.

Bostrom, Nick. 2011. Transhumanist versus bioconservatives. *H±Transhumanism and Its Critics* 25–29.

Butler, Octavia. 2014. *Parable of the talents*. London: Headline Publishing Group.

Duarte, Bárbara Nascimento. 2014. Entangled agencies: New individual practices of human-technology hybridism through body hacking. *NanoEthics* 8 (3): 275–285.

Dvorsky, George. 2015. This biohacker used eyedrops to give himself temporary night vision. *Gizmodo*, 27 March 2015. https://gizmodo.com/this-biohacker-used-eyedrops-to-give-himself-temporary-1694016390. Accessed 12 Aug 2021.

Edman, Timuçin Buğra. 2019. Transhumanism and singularity: A comparative analysis of a radical perspective in contemporary works. *Gaziantep University Journal of Social Sciences* 18 (1): 39–49.

Eglash, Ron. 2016. Of Marx and makers: An historical perspective on generative justice. *Teknokultura* 13 (1): 245–269.

Eglash, R., W. Babbitt, A. Bennett, K. Bennett, B. Callahan, J. Davis, J. Drazan, C. Hathaway, D. Hughes, M. Krishnamoorthy, and M. Lachney. 2017. Culturally situated design tools: Generative justice as a foundation for STEM diversity. In *Moving students of color from consumers to producers of technology*, 132–151. IGI Global.

Fukuyama, Francis. 2004. Transhumanism. *Foreign Policy* 144: 42–43.

Grewe-Salfeld, Mirjam. 2021. *Biohacking, bodies and do-it-yourself: The cultural politics of hacking life itself*, 314.

Hall, Melinda. 2016. *The bioethics of enhancement: Transhumanism, disability, and biopolitics.* Lexington Books.

Hester, Helen. 2018. *Xenofeminism*. Wiley.

Laboria Cuboniks. 2018, *The Xenofeminist manifesto: A politics for alienation.* Verso Trade.

Lee, James. 2015. *Biohacking manifesto: The scientific blueprint for a long, healthy and happy life using cutting edge anti-aging and neuroscience based hacks.*

Lee, Stephanie M. 2017. This guy says he's the first person to attempt editing his DNA with CRISPR. *Buzzfeed News*, 14 October 2017. https://www.buzzfeednews.com/article/stephaniemlee/this-bio hacker-wants-to-edit-his-own-dna#.evELlvD9p. Accessed 12 Aug 2021.

Manzocco, R. 2019. *Transhumanism, engineering the human condition.* Chicester, UK: Springer.

More, Max, and Natasha Vita-More, eds. 2013. *The transhumanist reader: Classical and contemporary essays on the science, technology, and philosophy of the human future.* West Sussex: Wiley.

Ming, Chong Lu. 2020. *Hacking voice assistants: Speculative design as resistance in the age of surveillance capitalism.* PhD diss., Massachusetts Institute of Technology.

Porter, Allen. 2017. Bioethics and transhumanism. *Journal of Medicine and Philosophy* 42 (3).

Trimper, John B., Paul Root Wolpe, and Karen S. Rommelfanger. 2014. When "I" becomes "We": Ethical implications of emerging brain-to-brain interfacing technologies. *Frontiers in Neuroengineering* 7:4.

Weise, Jillian. 2016. The dawn of the Tryborg. *New York Times*, 30 November 2016. https://www.nytimes.com/2016/11/30/opinion/the-dawn-of-the-tryborg.html. Accessed 4 Oct 2021.

Williams, Damien P. 2019. Heavenly bodies: Why it matters that cyborgs have always been about disability, mental health, and marginalization. *Mental Health, and Marginalization* (June 8, 2019).

Wohlsen, Marcus. 2012. *Biopunk: Solving biotech's biggest problems in kitchens and garages.* Current.

Dr. Emma Tumilty is feminist bioethicist and Assistant Professor at University of Texas Medical Branch Galveston. Her interests include matters of research ethics and innovation and health justice and access. She sees biohacking and related practices of open medicine, SynBio, biopunk practices, etc. as offering exciting opportunities for discussion around the "who does" and "how" of "science" and "medicine." Her interest in tranhumanism is focused on those rarely given the attention in mainstream settings—the crip technoscience writers, feminist technomaterialists, afro-futurists, etc. who imagine different communities, morphological freedoms, and futures.

Chapter 2
Pragmatism and Transhumanism

Aaron Bruce Wilson

Abstract In common parlance, the terms "pragmatism" and "pragmatic" signal a preference for practical expediency over a commitment to such lofty ideals and goals as the transhumanist goal of "the exploration of the posthuman realm" (Bostrom, Journal of Philosophical Research 30:3–14, 2005b). However, as embodied in the thought of its two great founders, Charles S. Peirce and William James, *philosophical* pragmatism seems to identify with such ideals and goals. Not only does Peirce identify our highest ends (truth and the ethical summmum bonum) with what would appear to be a posthuman state of knowledge and being, James's melioristic metaphysics conceptualizes reality itself as something "waiting to receive its final touches at our hands", which, on a cosmic scale, also seems to imply a posthuman state of being. While neither Peirce nor James envisioned the sorts of technologies that could allow for indefinite life extension, or radical physical or intellectual enhancement, both regarded science to have an unlimited potential to improve human conditions. Even later pragmatists, such as Richard Rorty, viewed science as being in the service of humanity and as having indefinite potential to improve human conditions.

2.1 Introduction

Within academic philosophy, transhumanist ideas receive the most attention from bioethicists, as proposed enhancement, life extension, and A.I. technologies abound in ethical questions. However, transhumanism is not simply a pro-technology movement that raises ethical questions. As found in the work of two leading transhumanist philosophers, Max More and Nick Bostrom, *transhumanism is a philosophical approach to fundamental issues*, metaphysical and metaethical. More and Bostrom have also recognized a rich history of transhumanist philosophy predating their own

A. B. Wilson (✉)
South Texas College, McAllen, US
e-mail: awilson3@southtexascollege.edu

E. Tumilty and M. Battle-Fisher (eds.), *Transhumanism: Entering an Era of Bodyhacking and Radical Human Modification*, The International Library of Bioethics 100,
https://doi.org/10.1007/978-3-031-14328-1_2

work,[1] beginning most ostensibly with renaissance and enlightenment thinkers such as Pico della Mirandola, Bacon, and Condorcet, but also including some nineteenth-century philosophers such as Nietzsche and Nikolai Federov (Federov's *cosmism* is perhaps the clearest philosophical precursor to transhumanism).[2] Moreover, as Bostrom notes, "[a]fter the publication of Darwin's *Origin of Species* (1859), it became increasingly plausible to view the current version of humanity not as the end point of evolution but rather as an early phase" (2005a, 3). With that in mind, perhaps no philosophical tradition is better able to absorb and provide theoretical motivations or support for transhumanist ideas than *pragmatism*.

Born from Darwinian naturalism and late-nineteenth-century scientific optimism, pragmatism takes a naturalistic, evolutionistic,[3] and melioristic[4] approach to human nature, experience, and society. It resists essentialist and dualistic views of humanity, and it argues for our ability to radically improve our conditions through democratic processes and the applications and advancements of the sciences. For the most part, pragmatism upholds natural science as our epistemic gold standard.[5] Yet, imbued with religious sentimentality,[6] pragmatism is also hospitable to the spiritual strains within transhumanist literature, including the idea of becoming "god-like" or merging with God through a self-guided evolution. But these connections between pragmatism and transhumanism are subtle enough that More and Bostrom miss them in their accounts of the history, and so I will elaborate on them here.

In common parlance, the terms "pragmatism" and "pragmatic" signal a preference for practical expediency over a commitment to lofty ideals and goals—where a "pragmatist" may even suggest the sort of person who would scoff at alleged transhumanist "pipe dreams" or "science fiction fantasies". But in philosophy, "pragmatism" most generally refers to an emphasis on connections between cognition (or "theory") and

---

[1] Bostrom (2005a); More (2013).

[2] Federov (1990/1906).

[3] For a focused account of the influences of Darwin's theory and other evolutionary hypotheses on the development of philosophical pragmatism, and on various early pragmatists, see Pearce (2020).

[4] It is generally agreed that the pragmatists, from Peirce to Rorty, held a melioristic ethics and, in some cases, metaphysics, viewing progress as being in some way essential to continued human existence. However, the nature of their meliorism is subject to different analyses. Recently, Koopman (2015) recognizes a strain of perfectionism in pragmatist's meliorism while Liszka (2021b) denies that strain of perfectionism and emphasizes a "tragic" aspect to it. According to Liszka, progress is not inevitable on the pragmatist's view, however, here I argue that for at least one pragmatist, Charles S. Peirce, progress is inevitable over the long run, on the condition of the continued of intelligent life.

[5] Note that some forms of pragmatism, notably Richard Rorty's, reject the epistemological concepts needed to make sense of an "epistemic gold standard" at all.

[6] Peirce, James, and Dewey all engaged with religion in a positive and constructive manner (Peirce with his "Neglected Argument for the Reality of God", James with works such as "The Will to Believe", and Dewey with "A Common Faith"). Slater (2015) is a good recent overview of their positive and constructive engagement with religion. However, whether their engagement with religion violates their naturalism is controversial. Slater (2015) argues that it is anti-naturalistic, while scholars such as Raposa (2018) argue that, at least in the case of Peirce, their engagement with religion incorporates a form of naturalism.

practice, between concepts such as *truth* and normative or goal-directed activities such as inquiry, communication, and political participation. I will argue here that its emphasis on the connections between the concept of *truth* and the practice of *inquiry* puts it in a primary and positive relation to transhumanism.

However, as with transhumanism, pragmatism is far from having been a mono-lithic philosophical movement. Just as common themes unify transhumanists more than any exact theories or agendas, so too the authors typically identified as pragma-tists are so associated more by common themes than by any exact theories or agendas. Among the post-War generation, authors as diverse as Cornel West, Susan Haack, and Robert Brandom have self-identified as some type of pragmatist, despite that their writings bear little resemblance to one another. While they all claim intellectual heritage from *some combination* of the "big three" classical pragmatists—Charles S. Peirce (1839–1914), William James (1842–1910), and John Dewey (1859–1952)— pragmatists today are markedly influenced by other intellectual figures, from W.E.B. Dubois to Rudolph Carnap. Most pragmatists active today probably have not paid much attention to transhumanism specifically. Some might possibly be hostile to transhumanism, while others seem dismissive of it. In his "Pragmatism and Death", Lachs (2014) suggests that it is "parody of pragmatism" to take pragmatists as "fore-seeing a time at which we will be essentially immortal" (379). I disagree; it is more likely a promise of pragmatism than a parody of it. At least, I argue that such tran-shumanist ideas find a home in the two self-identifying *progenitors* of pragmatism: Charles S. Peirce and William James.[7]

Peirce and James form the historical core of pragmatism. Peirce first articulates the "principle of pragmatism" in his seminal 1878 article "How to Make Our Ideas Clear", and in his 1898 UC Berkeley lecture, James first refers to the principle of pragmatism *as such*, crediting Peirce for having first articulated it. James did more than anyone else to popularize the idea that "meaning consists in practical effects". But Peirce considered the directions that James and others took this idea to be too literary (as opposed to scientific) in style, which prompted Peirce to return to and develop his pragmatism more systematically, and to rename it "pragmaticism"—a term he thought ugly enough "to be safe from kidnappers" (5.414, 1905).[8]

In my view, the strongest connections between pragmatism and transhumanism lie with Peirce rather than with James. I argue that Peirce's pragmatism directly implies a type of transhumanism, on two grounds. First, his pragmatist theory of *reality* as "the object" of "the opinion fated to be ultimately agreed to by all who investigate" implies that human inquirers also would (and should) eventually augment their cognitive

---

[7] Dewey occasionally identified as a "pragmatist" (e.g. Dewey 1908, 85) although it is unclear if or how his pragmatism is distinct from the views he adopted as "experimentalism", "instrumentalism", "empirical naturalism", and others. While Dewey was a pragmatist, I and others consider Peirce and James to be the original authors, while Dewey properly belongs to a *second* generation of pragmatists. See Legg and Hookway (2020).

[8] Most references to Peirce's writings are to *the Collected Papers of Charles S. Peirce* and are denoted with "CP" followed by the volume number and the paragraph number. See the works cited for full bibliographic information and for references to other collections of Peirce's writings and corresponding in-text abbreviations.

abilities to access realities to which current humans are cognitively closed. Second, his account of our ethical *summum bonum*, or ultimate good, which is also motivated by his pragmatism, is strikingly suggestive of the *transhumanist* summum bonum, best described by Bostrom (2005b) as "the exploration of the posthuman realm".

These connections between Peirce and transhumanism are explained with greater detail in Brunson and Wilson (2017). Here I will also explain how James's pragmatism offers rich support for certain transhumanist ideas, and how even the pragmatist who is most opposed to Peirce's form of pragmatism still holds a form of pragmatism that seems favorable to transhumanism: Richard Rorty (1931–2007). Rorty argues that we are not accountable to anything beyond other human beings, not even to an *objective reality*. He claims not so much to reject the notion of objectivity as he does to reject the very distinction between the objective and the subjective. Thus, he regards his own endorsement of the ideal of *global solidarity* to be neither relative nor objective, although, as an endorsement of a certain cultural value, he admits it is ethnocentric (Rorty 1990).

While I think Rorty's "ethnocentric" pragmatism is incoherent, in replacing the notion of objectivity with solidarity and hope in the continual improvement of the human community, as well as in upholding the scientific enterprise as a *model* of solidarity, Rorty's pragmatism is still well aligned intellectually with transhumanism. If I can show that even the two forms of pragmatism most opposed to each other, Peirce's and Rorty's, are well aligned with transhumanism, then that should take me far in showing that transhumanism and pragmatism, while certainly not inseparable, could each mutually benefit from the study of the other's ideas.

In what follows, first I explain how James's pragmatism and melioristic metaphysics (what I call "cosmic meliorism") supports transhumanist goals, and then I explain how Peirce's pragmatist accounts of truth and the ethical summum bonum direct us toward the ultimate goal of exploring the posthuman realm. Finally, I explain how even Rorty's pragmatism, with its anti-metaphysical meliorism, takes us in the direction of radical cognitive enhancements that might improve human conditions.

2.2   William James

A key transhumanist idea is that our role in the cosmos is not limited to our current modes of life on the surface of our home planet. Humans can play a greater role in the cosmos, likely transforming themselves in the process. From an anthropocentric perspective, the idea is that the cosmos is ready-made and just exists out there awaiting human exploitation. Reasonably, this would strike many as a predatory view of our place in the cosmos. However, from a certain metaphysical perspective, the idea is that the cosmos is *not* ready-made but is still in the process of creation, and that we play a constitutive role in that process. What exists *so far* awaits the guidance of intelligent beings for further development, and we cannot separate our interests, aims, and values from the cosmos. Far from being merely subjective, our interests, aims, and values form part of and develop with the cosmos. While this metaphysical perspective might elevate transhumanism above criticisms of unhinged cupidity and intellectual shallowness, how plausible is it? Why think that our interests and values should have any cosmic significance?

One reason might be that we really have no *concept* of the cosmos unshaped or uninfluenced by our interests and values, no matter how mathematically abstract that concept might be. A common argumentative strategy in metaphysics is to argue for a certain theory regarding our concepts and ideas, and then argue for a certain metaphysical thesis based on that theory of concepts.[9] Peirce's and James's pragmatism each pursue this strategy. They each put human interests front and center *metaphysically* by putting them front and center *conceptually*. Just as the *philosophical empiricist* (e.g., David Hume) argues that we have no concept of the world beyond our sensory experience, so the *philosophical pragmatist* argues that we have no concept of the world beyond our sensory experience *and* our interests. What we *mean* by "the world", "the cosmos" or "reality" is inevitably related to our interests and values in some form or other. So, on the (Peircean and Jamesian) pragmatist view, we try to make the best sense of reality or the cosmos *as* something related to our interests and values—minding that our interests include not merely the mundane and material but also those higher up on Maslow's hierarchy, such as freedom and self-actualization.

However, this point about pragmatism might come across more clearly in other works of James's than it does in his 1898 UC Berkeley lecture,[10] which is widely recognized as having begun the pragmatist movement in philosophy. There James introduces Peirce as "one of the most original or contemporary thinkers" and he introduces "Peirce's principle" or the principle of pragmatism as saying that "to attain perfect clearness in our thoughts of an object, then, we need only consider what practical effects the object may involve—what sensations we are to expect from it, and what reactions we must prepare" (1977, 348). Both in Peirce's original formulation and in James's reformulation of it here, it is not quite explicit that human interests or values play an essential role in our concepts of things. James's reformulation barely even distinguishes the principle of pragmatism from traditional empiricist principles that explain concepts in terms of sensory experiences. James almost admits as much:

I am happy to say that it is the English-speaking philosophers who first introduced the custom of interpreting the meaning of conceptions by asking what difference they make for life. Mr. Peirce has only expressed in the form of an explicit maxim what their sense for reality led them all instinctively to do. The great English way of investigating a conception is to ask right off, "What is it known as? In what facts does it result? What is its cash-value, in terms of particular experience? (1977, 360).

---

[9] For example, in his famous *Meditations on First Philosophy*, Rene Descartes argue for substance dualism about mind and body based on our conceptions of each.

[10] This lecture was delivered to the Philosophical Union of the University of California on August 26, 1898. It was reprinted in James's 1907 Pragmatism as "Philosophical Conceptions and Practical Results" (p.p 350–355) and in *The Writings of William James: A Comprehensive Edition* pp. 345–361. References to James are to this comprehensive edition. See works cited for full bibliographic information.

While James's promotion of Peirce in the 1898 UC lecture helped revitalize the latter's philosophical work, Peirce comes to distance himself from James's formulation of the pragmatist principle. One reason is James's emphasis on "particular" experiences. For Peirce, meaning consists not in particular experiences but in *types* of experiences that one *would have* under certain *types* of conditions. Peirce consistently rejects what he calls nominalism in metaphysics, which emphasizes particulars over generals and tokens over types. He found not only James's pragmatism but even his own original 1878 formulation of the pragmatist principle to have gone too far in the direction of nominalism (CP 8.208, 1905).

Moreover, starting around 1898, and likely also in response to James, Peirce begins to emphasize how, for pragmatism, interests or purposes, and not *just* sensory experiences, play a constitutive role in our conceptual representations. Around 1898, he develops the idea that concepts are formed by associations between sensations and *interests*, which he calls "association by interest" (CP 7.499), and in his 1903 Harvard lectures he concludes that, upon pragmatism, "[t]he elements of every concept enter into logical thought at the gate of perception and make their exit at the gate of *purposive action*; and whatever cannot show its passports at both those two gates is to be arrested as unauthorized by reason" (CP 5.212).[11] Further, two years later, Peirce remarks that "the preference for the name pragmatism" derives from "its recognition of an inseparable connection between rational cognition and rational *purpose*" (EP 2:333/Peirce 1998).

That James's pragmatism also regards interests and purposes as essential to our concepts is clearer in works published both before and after the famous 1898 UC lecture. Nearly twenty years earlier, in *The Sentiment of Rationality* (1879), James argues: "Conceptions, kinds, are teleological instruments. No abstract conception can be a valid substitute for a concrete reality except with reference to a particular interest in the conceiver" (321). The reason is that cognition in general, on his view, "is but a fleeting moment, a cross-section at a certain point, of what in its totality is a motor phenomenon", so that "[c]ognition, in short, is incomplete until discharged in act" (330). James makes the Darwinian naturalist argument that human cognition should be no different in its basic nature from that of less complex life, noting that "[i]n lower forms of life no one will pretend that cognition is anything more than a guide to appropriate action" (ibid.).

However, in his 1906 lectures of pragmatism, James seems to take the idea that interests play a constitutive role in our concepts further than Peirce ever had, to the point where the satisfaction of certain interests is *all* that matters for the validity of a concept. In the second lecture, "What Pragmatism Means", James presents pragmatism as a radical type of empiricism, one that regards concepts and theories as "instruments" functioning, *not* to represent an independent reality, but to connect our experiences in ways that satisfy certain interests. Thus, as James presents it, the

---

[11] In these lectures, Peirce articulates his pragmatism in terms of the "normative sciences", writing: "as pragmatism teaches us, what we think is to be interpreted in terms of what we are prepared to do, then surely *Logic,* or the doctrine of what we ought to think, must be an application of the doctrine of what we deliberately choose to do, which is Ethics" (CP 5.35).

"pragmatist theory of *truth*" holds that a theory is true so far as it satisfies certain interests, such as "linking things satisfactorily, working securely, simplifying, saving labor" (382). While Peirce does not directly criticize James for reducing truth to the satisfaction of human interests—an extreme to which Peirce never takes his own pragmatism—he criticizes James for being too unclear regarding the exact interests that theories need to satisfy to be true. Peirce remarks: "Mr. Ferdinand C.S. Schiller informs us that he and James have made up their minds that the *true* is simply the *satisfactory*. No doubt; but to say "satisfactory" is not to complete any predicate whatever.

Satisfactory to what end?" (CP 5.554, 1906). As I will explain soon, Peirce answers this question quite precisely, and without abandoning a representational view of truth.[12]

The final three lectures of James's 1906 pragmatism series explain how (his) pragmatism implies *meliorism*,[13] and here I think lies the strongest connection between Jamesian pragmatism and transhumanism. As James explains it, while *optimism* views "the world's salvation" as inevitable, and while *pessimism* views it as impossible, *meliorism* "treats it as a possibility which becomes more and more of a probability the more numerous the actual conditions of salvation become" (1977, 466–67). James notes that we may "interpret the word 'salvation' in any way [we] like", but that it is clear that "pragmatism must incline towards meliorism" (1977, 467).

The reason pragmatism must incline toward meliorism stems from James's pragmatist view of truth. On that view, truth consists in "verification processes" themselves, so that "truth is made" and does not "absolutely obtain" the moment a proposition is thought or uttered, as it does on the "rationalist" view of truth (436–438). James argues that, for the rationalist, "[r]eality stands complete and ready-made from all eternity," such that "the agreement of our ideas with it is [a] unique unanalyzable virtue in them" (439). However, the Jamesian pragmatist cannot accept such a view of reality, and so he cannot accept the rationalist view of truth as "inert, static, a reflexion merely". "For pragmatism," James writes, "[reality] is still in the making, and awaits part of its complexion from the future" (456). The Jamesian pragmatist views the world as "incomplete" or "in process"; so, she views truth or the relation between thought and the world as also being "in process".

But why must the (Jamesian) pragmatist hold that reality is in process or "still in the making"? —because we have no conception of reality separate from our purposes and interests and we always have some purpose or interest that *we are in the process of satisfying*. As, according to pragmatism, concepts are structured by and make implicit reference to human experiences and interests, so it is impossible "to separate the real from the human factors in the growth of our cognitive experience" (1977, 454). James offers some examples, including the carving out of groups of stars into constellations,

---

[12] Here "representationalism" refers to the view that our thoughts, sentences, and theories function to represent the world. "Anti-representationalism" holds that these thoughts, sentences, and theories do not represent the world but function only to direct us toward the satisfaction of certain ends.

[13] Lecture VI: Pragmatism's Conception of Truth, Lecture VII: Pragmatism and Humanism, and Lecture VIII: Pragmatism and Religion.

but it might not be too difficult to show that human interests also shape the character of our mathematical representations of reality. No matter how abstract the theory, James holds that "you can't weed out the human contribution" (455). At least, *we implicitly conceive reality as a field or space of potentials for the satisfaction and growth of our interests and ends.* It is not unlike Gibsonian "affordances" but applied to the conceptual as well as the perceptual level of cognition. We inevitably conceive reality in relation to ourselves: as something we can reshape and that can also reshape us. "The world stands really malleable", James writes, "waiting to receive its final touches at our hands" (1977, 456).

So, pragmatism "must incline toward" meliorism because pragmatism holds, first, that we *cannot* attain a clear concept of the world as complete and favorable to us (optimism) or as complete but unfavorable to us (pessimism), but that, and second, we can attain a clear concept of the world as *incomplete* and favorable or unfavorable to us *depending on our own actions* (meliorism).

Thus, in James, meliorism is not simply a social or ethical outlook, it is metaphysical thesis: we have no concept of the world except as something that can be molded in relation to our interests. That is, in relation to our interests, the world is essentially something "resisting yet malleable" (1977, 456).[14] James does not minimize the "resisting" aspect of reality (an aspect that Peirce strongly emphasizes) but he insists that its malleability is the only way it can appear *rational* to us. He remarks that "the only fully rational world would be the world of wishing-caps, where every desire is fulfilled instanter, without having to consider or placate surrounding or intermediate powers" (1977, 268). Our drive to make rational sense of the cosmos is a drive to reshape it according to our interests. As he sees it, *theism* is the attempt to make rational sense of the world by positing a God who is directly concerned with and favorable to our interests. James concludes his 1906 lectures on pragmatism by suggesting that, in fact, God is a *melioristic ideal.* He writes:

I firmly disbelieve, myself, that our human experience is the highest form of experience extant in the universe. I believe rather that we stand in much the same relation to the whole of the universe as our canine and feline pets do to the whole of human life. …But, just as many of the dog's and cat's ideals coincide with our ideals, and the dogs and cats have daily living proof of the fact, so we may believe … that higher powers exist and are at work to save the world on ideal lines similar to our own. …You see that pragmatism can be called religious, if you allow that religion can be pluralistic or merely melioristic in type. (1977, 472).

Not surprisingly, meliorism is a prominent theme in other works by James and, in those works too, it has a transhumanist appeal. For instance, in 1910's "The Moral Equivalent of War", James responds to the argument that peace between nations is not sustainable because militaristic discipline and cooperation are necessary to sustain nations and prevent their degeneration. He argues that militaristic values can be sustained peacefully with an "army enlisted against *Nature*" to improve human conditions. Although the "army" James has in mind seems mostly suited for hard

---

[14] Koopman (2006, 107) offers a similar account of the metaphysical aspects of James's meliorism. Slater (2015) among others also recognize James's meliorism as a metaphysical viewpoint.

labor rather than scientific and technological advancement (1977, 669), the idea of an "army enlisted against *Nature*" can easily be taken in the latter (transhumanist) direction.

The Jamesian pragmatist view that our theories or concepts of Nature inherently relate to our interests is favorable to the transhumanist idea of shaping Nature, including human nature, to suit our ultimate interests. Pragmatism argues that, in knowing Nature, we are not merely spectators of it but also active participants in its evolution. Knowledge itself constitutes rules for subjecting Nature to our will *and* for reshaping our will to Nature. Our very concept of reality is geared towards a kind of cosmic meliorism: to be real is to play a role in the satisfaction or the development of human interests. And insofar as radical technological advancements, such as the artificial augmentation of human capacities, are necessary for and are actually geared toward this cosmic meliorism, Jamesian pragmatism would support such advancements.

2.3 Charles S. Peirce

Just as James's pragmatism grounds an account of truth that is favorable to transhumanist ideas, so Peirce's pragmatism grounds an account of truth that is favorable to transhumanist ideas. However, they are favorable to transhumanist ideas in different ways. While James's pragmatist theory of truth, as "being made" as opposed to "absolutely obtaining", implies a melioristic metaphysics that indirectly supports such projects as augmenting human capacities and extending human life spans, Peirce's pragmatist theory of truth supports the project of augmenting human capacities more directly by implying that our very commitment to *truth* itself requires that we augment our cognitive capacities—that is, *if* there are any realities to which current humans are cognitively closed. And the idea that there are realities or modes of being that are inaccessible to current humans because of their biological limitations is a common theme among transhumanists, exemplified prominently in Bostrom (2005b).

While Peirce repeatedly denies, on pragmatistic grounds, that there could be any unknowable or incognizable realities, where he does so he means realities that are incognizable or unknowable *by any possible knower or inquirer*, not just by current human beings. In fact, he argues that our very concept of *reality* "essentially involves the notion of a community, without definite limits" (CP 5.311, 1868), adding later that this community "must not be limited, but must extend to all races of beings with whom we can come into immediate or mediate intellectual relation" (CP 2.652, 1878). So although Peirce does not assume that current humans *must* augment their cognitive abilities to access realities that are closed to them, he assumes that, *if* there are realities unknowable to current humans, then there at least needs to be members of our "intellectual community" to whom those realities *are* knowable. Otherwise, the responsibility would be on us either to create such members (e.g., A.I. +) or to artificially augment our own capacities.

But why does Peirce's pragmatist account of truth imply that, if there are realities that are unknowable to current humans, then we should either augment our cognitive capacities or create creatures for whom those realities are knowable? Here I need to explain that account truth.

It is in "How to Make Our Ideas Clear" (1878) that Peirce first articulates "the rule for attaining the third grade of clearness in our concepts"—i.e., the principle of pragmatism, or what came to be known as "the pragmatic maxim". The "third grade of clearness" is supposed to render our concepts even clearer than *abstract definitions* do, which afford only a "second grade of clearness" by articulating the necessary and sufficient conditions for a certain object to be signified by a certain concept. The third grade, in contrast, involves grasping "the practical effects" of the type of object signified by the concept. With the concept of *reality*, Peirce argues that the most general practical effect of a *real* object is that *it tends to fix our belief in it*. While a real object might escape our recognition for an exceptionally long time (say, a speck of dust a billion light years away), if it would have any practical effect at all, it would be to, eventually, fix our belief in it. Thus, while we can abstractly define "truth" in terms of an abstract correspondence between a proposition and reality, pragmatically, "the opinion which is fated to be ultimately agreed to by all who investigate, is what we mean by *the truth*, and the object represented in this opinion is *the real*" (CP 5.407, 1878; my emphasis).

Notice that Peirce formulates his pragmatic clarification of *truth* in direct connection with his pragmatic clarification of *reality*. At the "pragmatic" grade of clearness, the concepts of truth and reality are invariably linked. However, the link between pragmatism and these notions of truth and reality, as the "final opinion" and the object represented in that opinion, is actually post hoc, as they first appear seven years earlier in his 1871 review of Fraser's edition of *The Works of George Berkeley*.[15] There, Peirce seems to base them, instead, on empirical hypotheses such as that "human opinion universally tends in the long run to a definite form" (CP 8.12).[16] Generally, it may seem that this notion of truth makes a strong empirical prediction: that a "final opinion" *will* be reached on all matters concerning reality. However, he clarifies later that the final opinion only *would* be reached *if* the community of inquirers *were* to survive long enough; that is, there is a *real tendency toward this final opinion*, so that it would be reached given enough time—although it might not

---

[15] In that 1871 review, Peirce argues: All human thought and opinion contains an arbitrary, accidental element, dependent on the limitations in circumstances, power, and bent of the individual; an element of error, in short. *But human opinion universally tends in the long run to a definite form, which is the truth.* Let any human being have enough information and exert enough thought upon any question, and the result will be that he will arrive at a certain definite conclusion, which is the same that any other mind will reach under sufficiently favorable circumstances. ... There is, then, to every question a true answer, a final conclusion, to which the opinion of every man is constantly gravitating. He may for a time recede from it, but give him more experience and time for consideration, and he will finally approach it. ... This final opinion, then, is independent, not indeed of thought in general, but of all that is arbitrary and individual in thought; is quite independent of how you, or I, or any number of men think. *Everything, therefore, which will be thought to exist in the final opinion is real, and nothing else.* CP 8.12 (my emphasis).

[16] See Wilson (2018)

2 Pragmatism and Transhumanism

actually be reached because, for instance, there might not *actually* be enough time. While this still makes an empirical prediction, it is statement of a *ceteris paribus law* making only a conditional prediction rather than an absolute one.

Nevertheless, why suppose even this conditional prediction is correct? If the "final opinion" is supposed to be a final opinion on *all* reality, it might *never* be reached simply because there may be realities we could never know. This objection, that there may be realities that *no* amount of inquiry could settle our opinions about, is known by scholars as "the problem of buried secrets". In the seminal 1878 paper, Peirce responds to this objection as follows:

[I]t is unphilosophical to suppose that, with regard to any given question (which has any clear meaning), investigation would not bring forth a solution of it, if it were carried far enough. Who would have said, a few years ago, that we could ever know of what substances stars are made whose light may have been longer in reaching us than the human race has existed? Who can be sure of what we shall not know in a few hundred years? Who can guess what would be the result of continuing the pursuit of science for ten thousand years, with the activity of the last hundred? And if it were to go on for a million, or a billion, or any number of years you please, how is it possible to say that there is any question which might not ultimately be solved? (CP 5.409).

Such speculative optimism should resonate with transhumanists, but it does not really answer the objection. Elsewhere Peirce suggests that questions that would never get fixed answers simply have no *true* answers, such that there would be "a lacuna in the completeness of reality" (CP 8.156, c.1900). But to answer this objection, Peirce could simply retreat to the *second grade of clearness* and say that, while the assumption that there are unknowable realities can be abstractly understood, it simply cannot be made pragmatically clear.[17] As Peirce explains in a later writing, the third (pragmatic) grade of clearness does not supersede the lower grades of clearness, such as abstract definitions (CP 8.218, 1910). Concepts grasped *only* at those lower grades of clearness are still meaningful in some way,[18] although he insists that ideas that are comprehensible only at the second or abstract grade are virtually meaningless to empirical inquiry and can even "block the way" of it (CP 1.138, 1898).

Peirce will also insist that the basic motivating *hope* of all inquiry is that there is one answer to which inquiry on any meaningful question would eventually converge (e.g., CP 5.407, 1878). Truth is the motivating hope of all inquiry because, according to him, truth is the ultimate *logical ideal*, the ultimate form of *logical goodness*, where logic is the normative science concerning "what we ought to think" (CP 5.35, 1903). This means that truth as the *would-be* final opinion is not only a state that we would approach given enough time, it is a state that we *should* approach, so far as doing so does not violate other (non-logical) ideals. On Peirce's view, we have a prima facie duty to pursue the truth, pragmatistically understood as the ultimate final opinion.

---

[17] See Wilson (2020). For more on the buried secrets objection, including alternative solutions to it available to Peirce, see Lane (2018), 60–67.

[18] See Wilson (2020).

Now, it is not that, for Peirce, we have a duty to pursue the final opinion on each meaningful question simply for the sake of each meaningful question, including questions like "how many grains of sand were on Earth on Jan. 2nd 1 million B.C.?" It is that we cannot pursue the final opinion on *any* meaningful question without pursuing the final opinion on *all* of them. This point is not so explicit in Peirce, and at some places he implies that the final opinion could be reached on some questions without being reached on others. However, it is impossible to approach the final opinion on some questions but not others, because, for Peirce, the final opinion is a point of *permanently fixed belief*, at which some unexpected experience that could shake up one's belief becomes impossible: all expectations or predictions are fulfilled. Because one can have a true belief and yet come to doubt it due simply to misleading circumstances resulting from ignorance about other facts,[19] to completely avoid all doubt on *any* matter one seems to require perfect knowledge on *all matters* (CP 4.62, 1893).

Thus, truth as the final opinion is best understood as a single comprehensive (albeit hypothetical) point that we would indefinitely approach, if not fully reach, if the community of inquirers had an indefinite amount of time. It is a state of virtual *omniscience*.[20] It is fair to say that, in Peirce, to pursue the truth is to pursue omniscience, to uphold truth as an *ideal* is to uphold omniscience as an ideal, and to say that some proposition *is* true is to say that it would be held in a state of omniscience.

So, there are at least two significant connections between transhumanism and Peirce's pragmatist account of truth. First, his account offers transhumanism a *conceptual argument* for radical cognitive enhancement. On the assumption that there are realities unknowable to current humans but still knowable in principle (realities *un*knowable in principle being mere abstractions), then it follows from his account of truth *and* his upholding truth as our ultimate logical ideal that we have a logical prima facie duty to try to overcome any cognitive limitations that might block our approach toward the final opinion.[21] It is a "prima facie" duty because it can be overridden by duties to pursue non-logical but *ethical* ideals, which Peirce regards as even more fundamental (read further on). Second, truth as the final opinion is a *posthuman state* of virtual omniscience, either as a sort of superorganism (a "community of inquirers" who collectively have ceased any need for further inquiry) or as a single mind into which all separate minds have gradually merged. While current humans can have true beliefs, the clearest sense we can make of these truths is in terms of what would be upheld by an evolved posthuman mind. And, so, to uphold the truth as an ideal is to uphold a posthuman state of cognition as an ideal.

Peirce's pragmatism aligns with Bostrom's (2005b) thesis that the *core* transhumanist ideal is "the exploration of the posthuman realm", not only with respect to the concept of truth but also with respect to the ultimate ethical ideal, the *summum*

---

[19] For example, if I believe truly that *P*, but a fried I trust lies to me out of an common prank, causing me to doubt that *P*, I will have doubted a true belief due my ignorance that the friend is lying to me.

[20] See Wilson (2016), 244–260, for my full explanation of Peirce's account of truth.

[21] See Brunson and Wilson (2017) for a fuller explanation of this point.

bonum. Peirce's writings on ethics and on the ethical summum bonum are more scattered, scant, and unclear than his writings on logic and truth. Nonetheless, significant philosophical conclusions can be drawn from them.

They are also connected into his pragmatism, for, as he says, "as pragmatism teaches us, what we think is to be interpreted in terms of what we are prepared to do, then surely logic, or the doctrine of what we ought to think, must be an application of the doctrine of what we deliberately choose to do, which is *Ethics*" (CP 5.35, 1903; my emphasis). As Peirce appears to argue in his 1903 Harvard lectures, the "practical effects" into which pragmatism resolves conceptual meaning concern not only our everyday interests, but also our *ultimate* interests, including truth and our ultimate ethical ideal. And one consistent interpretation of what Peirce believes our ultimate ethical idea to be, is the exploration of the posthuman realm.

Different interpretations of Peirce's writings on the ethical summum bonum are inevitable because they are notoriously cryptic.[22] He writes that our summum bonum is to contribute to "the process of evolution whereby the existent comes more and more to embody generals" (CP 5.433, 1905)—an "evolution" that scholars identify with what he elsewhere calls "the development of concrete reasonableness" (CP 5.3, 1902). For Peirce, "reasonableness" is the sort of lawful or organized complexity (what he calls "thirdness") that is most characteristic of *minds*. The "development of concrete reasonableness" refers to the development of the sort of organized complexity that not only would make the cosmos more comprehensible to intelligent beings, but would also expand the presence and the power of intelligence itself. It is the idea that "reasonableness", and intelligence as a form of "reasonableness", becomes an increasingly dominant characteristic of the cosmos. Moreover, Peirce says that later stages of this evolution or development takes "place more and more largely through self-control", meaning that intelligent beings themselves become increasingly responsible for the continued growth of reasonableness or intelligence across the cosmos.

Insofar as the growth of concrete reasonableness involves the growth of intelligence, it seems to correspond to the growth or approach toward "the final opinion" that Peirce postulates as our highest logical ideal. Certainly, to approach a point of virtual omniscience requires an overall growth of intelligence. Peirce even explicitly connects this idea with his *theism*, or belief in God. He argues that "I look upon creation as [still] going on and I believe that such vague ideas as we can have of the power of creation is best identified with the idea of *theism*", where "the [ultimate] ideal would be to fulfill our appropriate offices in the work of creation" (CP 8.138fn4). And "the purpose of creation", he argues, "as it must appear to us in our highest approaches to an understanding of it, is *to make an answering mind. It is God's movement toward self-reproduction*" (R 1334, 1905; my emphasis). In some writings Peirce seems committed to belief in the traditional God of creation, as opposed to the Jamesian "melioristic ideal"; but here we find that, for Peirce, the God of creation is entangled with a God that "self-reproduces"—a transcendent God

---

[22] For instance, Liskza (2021a, 64–47) offers an alternative account of Peirce's concept of the ethical summum bonum as "reasonableness" itself.

that is (was, will be) itself born originally from non-intelligent natural processes and evolved through a growth of "concrete reasonableness" could assume the role of the creator of those very natural processes, forming a sort temporal-causal loop. This might not be exactly what Peirce has in mind, but he seems to want to reconcile "God the creator" or traditional theism with God as the melioristic ideal and as a posthuman or self-evolved superintelligence.

Thus, Peirce's pragmatism can be accurately interpreted as a transhumanist philosophy, seeing that it implies that our ultimate logical and ethical ideal is the pursuit of a posthuman state of cognition and being. Most Peirce scholars have not explicitly recognized this because, first, transhumanism has not generally been on our radar and, second, his remarks on the possibilities and promises of technology are easily overlooked. But those remarks do appear, and they include speculations on the possibility of *thinking machines*. He remarks: "Precisely how much of the business of thinking a machine could possibly be made to perform, and what part of it must be left to a living mind, is a question not without conceivable practical importance" (NEM 3:625–32/ Peirce 1976). Recognizing how modern (circa 1900) technology "has put [humans] into quite another world, almost as much so as if it had transported our race to another planet" (CP 5.513, 1905), Peirce speculates that we might, with time, develop technologies allowing us to "find that the sound waves of Aristotle's voice have somehow recorded themselves" (CP 5.542, 1902). As a scientist himself, Peirce would be impressed but hardly surprised with the technological development since his time, and he would join transhumanists in pushing for its continued but cautious and ethical development.

2.4   Richard Rorty

While Peirce and James constitute the historical and intellectual core of pragmatism as a philosophical tradition, pragmatism *generally* is hardly hospitable to transhumanist ideas unless pragmatism *as it has evolved* since Peirce and James is also hospitable to transhumanist ideas. Now, much of what goes under "pragmatism" today may be less hospitable or simply less relevant to transhumanism, but that would not be the result of philosophers carrying out Peirce's and James's thought to their ultimate conclusions. After Dewey, or after the height of Dewey's work in the 1920s and 30s, what was anything like a pragmatist "tradition" or "school of thought" arguably receded, blurred, and splintered. While some philosophers followed in the footsteps of Peirce, James and Dewey, such as Charles Morris (1901–1979) with his work in (Peircean) semiotics, what survived *directly* from the "big three" was most notably the scholarly study of their works.[23] What emerged as "pragmatism"

---

[23] For instance, the creation of the *Collected Papers of Charles S. Peirce* undertaken by Charles Hartshorne and Paul Weiss starting in 1925. The publication of the first two volumes in 1931 effectively created "Peirce scholarship", with Justus Buchler's *Charles Peirce's Empiricism* (1939) being the first notable book on Peirce's philosophy. Such interest in Peirce led to the creation of

by the 1950s were mostly reactions to logical empiricism within the analytic tradition,[24] such as the reactions we find in Quine, Sellars, and the later Wittgenstein, who (perhaps with the exception of Sellars) seem to have barely read Peirce, James, or Dewey. It also became common around that time to begin including logical empiricists such as Carnap and C. I. Lewis as pragmatists, as Quine casted Peirce's principle of pragmatism as a version of the verification principle of logical empiricism.[25]

This analytic and anti-verificationist trend of "pragmatism"—often referred as "neopragmatism"—can be said to have culminated in the work of Richard Rorty (1931–2007), who perhaps is the best-known pragmatist of the last 50 years and who presents himself as following in the footsteps of James and Dewey (but not so much Peirce). In *Consequences of Pragmatism*, Rorty (1982) remarks.

On my view, James and Dewey were not only waiting at the end of the dialectical road which analytic philosophy travelled, but are waiting at the end of the road which, for example, Foucault and Deleuze are currently travelling. (xviii).

On Rorty's reading, James and Dewey had fundamentally rejected the epistemological paradigm of philosophy that, on some accounts, began with Descartes, continued with Kant and post-Kantian philosophy, and then culminated in twentieth century analytic philosophy—a paradigm that Rorty characterizes as "the attempt to say 'how language relates to the world' by saying what makes certain sentences true" (xix). He thinks that this attempt is futile because, as he sees it, it treats language (or thought) as an extra-wordly thing that holds some special relation to the world, where in fact language is just another feature of the world. He argues that this is the upshot of James's account of truth: "On James's view, 'true' resembles 'good' or 'rational' in being a normative notion, a compliment paid to sentences that seem to be paying their way and that fit in with other sentences which are doing so" (xxv). As we have seen, this is not entirely inaccurate. James argues against the correspondence view of truth, on which truth "absolutely obtains", and he writes of concepts as instruments and of truths as *instruments that work*. For James, truth is not extra-worldly by any means.

However, as we have seen, while James thinks that there are substantial metaphysical implications to this view of truth, as being "in process" and not "absolutely obtaining", Rorty does not. In fact, Rorty eschews metaphysics as strongly as the logical verificationism that he rejects. Although both Peirce and James presented

---

the Charles S. Peirce Society in 1945 and to the quarterly journal, *The Transactions of the Charles S. Peirce Society* in 1965, which continues to be the premier outlet for scholarship on classical pragmatism.

[24] The "analytic tradition" is thought by many to have begun with Gottlob Frege (1948–1925) and his 1892 paper "Sense and Reference" and with Bertrand Russell (1872–1970) and his 1905 paper "On Denoting". These paper ushered in the "linguistic turn" that is generally considered a dominant feature of early analytic philosophy.

[25] See Quine (1981), 30. The "Verification Principle" or the "Principle of Verifiability" holds that the meaning of sentences consists only their empirically verifiable truth conditions. Peirce's principle of pragmatism, in contrast, does not limit their meaning to empirically verifiable truth conditions, but to any "practical consequences", including those that might not quite "verify" the sentence. For instance, "The plate is hot" is not verified by "I would hold the plate with bare hands".

pragmatism as a method of ridding philosophy of meaningless metaphysics, they hardly thought *all* metaphysics to be meaningless. Moreover, the emphasis that Rorty gives to *language* relates him more to traditional analytic philosophy than to James or to Dewey, in whose works language per se is not nearly as prominent a subject as psychological notions such as thought or belief. Even in Peirce, the most "analytic" of the three, language is not nearly emphasized as much as is the broader concept of *signs*. But Peirce's and James's forms of pragmatism are fundamentally concerned with *truth*, and it is James's theory of truth (though not Peirce's) that Rorty finds important enough to adopt the title of "pragmatism".[26]

Rorty follows James not only in his view of truth as "what works" but also in adopting a melioristic view of the world. But in complete opposite fashion to James's metaphysical meliorism, Rorty's meliorism is motivated by a kind of *anti-metaphysics*. Rorty argues for further secularization and democratization of society, not necessarily because these are goods in themselves, but because they are conducive to achieving *global solidarity*. His anti-metaphysics comes down to the contention that we should "drop the idea that human beings are responsible to a nonhuman power" (1990, 39), whether that non-human power is a *God* or an "objective" world conceived as existing independently of human language, experience, and interests. Rorty thinks that if we drop this idea, then the proper focus will be on *improving* ourselves and our societies. He writes:

[The Pragmatist] is suggesting that instead of invoking anything like the idea-fact, or language-fact, or mind-world, or subject-object distinctions to explicate our intuition that there is something out there to be responsible to, we just drop that intuition. We should drop it in favor of the thought that we might be better than we presently are—in the sense of being better scientific theorists, or citizens, or friends. *The backup for this intuition would be the actual or imagined existence of other human beings who were already better (utopian fantasies, or actual experience, of superior individuals or societies)*. On this account, to be responsible is a matter of what Peirce called 'contrite fallibilism' rather than of respect for something beyond. The desire for 'objectivity' boils down to a desire to acquire beliefs which will eventually receive unforced agreement in the course of a free and open encounter with people holding other beliefs (Rorty 1990, 41; my emphasis).

Here we find that, in Rorty's view, "utopian fantasies" of "superior individuals or societies" play a positive role philosophically, directing our attention to the only thing that matters philosophically: the human community. While he does not say that these utopian fantasies need to be *techno-fantasies*, or that the "superior individuals" need to be artificially enhanced or *transhumans*, he does not say anything that would exclude them. In fact, even if our desire with respect to knowledge is not to attain absolute certainty or omniscience regarding a mind-independent reality, but to receive "unforced agreement" in the course of "free and open encounters" with others, various kinds of artificial enhancements would seem conducive to that goal. If we can edit our genes to increase our empathy, imagination, and other cognitive skills related

---

[26] Rorty famously but incorrectly remarks that Peirce's "contribution to pragmatism was merely to have given it a name, and to have stimulated James" (1982, 161).

to communication and socialization, we could maximize the potential to achieve unforced agreement among human (or transhuman) community.

Now, Rorty's vision of an "unforced agreement" among humanity is not like Peirce's vision of a "final opinion", as not only does Rorty think there is no need to view our unforced agreement as corresponding to a mind-independent reality, but also his vision does not involve everyone holding all the exact same beliefs or opinions. He explains that "pragmatists interpret the goal of inquiry (in any sphere of culture) as the attainment of an appropriate mixture of unforced agreement with tolerant disagreement where what counts as appropriate is determined, within that sphere, by trial and error" (1990, 41–42). Tolerance of differences is part of the goal of global solidarity, which aligns with the transhumanist value of morphological freedom: i.e., the freedom to change oneself physically as one sees fit. According to Rorty, even healthy competition has a place in global solidarity, so far as that competition enriches human activity. He writes:

We cannot, I think, imagine a moment at which the human race could settle back and say, 'Well, now that we've finally arrived at the Truth we can relax.' We should relish the thought that the sciences as well as the arts will always provide a spectacle of fierce competition between alternative theories, movements, and schools. The end of human activity is not rest, but rather richer and better human activity" (1990, 39).

While Rorty is not so naïve as to think that political competition never infects science, the value that he sees in science, beyond enriching human conversations and enabling technological advancements, is that it's a "model of human solidarity". It mixes intellectual competition that "enriches the conversation" in non-violent ways (at least, generally) with fundamental cooperation towards common goals.

For Rorty, science is in the service of human individuals and human societies. It is not in the service of an objective reality of which we are obligated to acquire a "spectator" type of knowledge—a type of knowing that is separable from our practical interests. Although many transhumanists and pragmatists (notably, Peirce) would disagree with it, such a conception of science focuses the goals of science in the direction of finding applications that would improve human existence. For Rorty, science is a means to an end, a human end, and the same general end for which transhumanists take a special interest in science: *the solution to fundamental human problems.*

I am unsure what most of my fellow students of pragmatism today think or would think of its connections with transhumanism. We are a diverse group, and I have had discussions on this only with a few fellow Peircean pragmatists. But of that sample, there is some positive reception. This paper serves as an invitation to researchers with interests either in pragmatism or in transhumanism to consider the parallels more deeply. Transhumanist ideas offer a fertile direction to take pragmatist lines of thought, and some core pragmatist ideas offer a rich philosophical basis for understanding and justifying transhumanist contentions. *Fundamentally*, why should we technologically pursue goals like radical cognitive enhancement, indefinite life spans, and the exploration of the posthuman realm? Transhumanism should have a metaphysics and ethics to support its prescriptive claims. I think that pragmatists, and especially Peirce, can provide that.

References

Bostrom, Nick. 2005a. A history of transhumanist thought. *Journal of Evolution and Technology* 14 (1): 1–25.

Bostrom, Nick. 2005b. Transhumanist values. *Journal of Philosophical Research* 30 (Supplement): 3–14.

Brunson, Daniel J., and Aaron B. Wilson. 2017. The transhumanist philosophy of Charles Sanders Peirce. *Journal of Evolution and Technology* 27 (2): 12–29.

Dewey, John. 1908. What pragmatism means by practical. *The Journal of Philosophy, Psychology and Scientific Methods* 5 (4): 85–99.

Fedorov, Nikolaï F. 1990. *What was man created for?: The philosophy of the common task*. London: Haymarket Publishing. Originally published vol. I, Verny, 1906, and vol. II, Moscow, 1913.

James, William. 1977. *The writings of William James: A comprehensive edition*, ed. John J. McDermott. Chicago: University of Chicago Press.

Koopman, Colin. 2006. Pragmatism as a philosophy of hope: Emerson, James, Dewey Rorty. *The Journal of Speculative Philosophy* 20 (2): 106–116.

Koopman, Colin. 2015. *Pragmatism as transition*. New York: Columbia University Press.

Lachs, John. 2014. Pragmatism and Death. In *Freedom and Limits*, eds. J. Lachs and P. Shade, 377–386. New York, NY: Fordham University Press

Lane, Robert. 2018. *Peirce on realism and idealism*. New York, NY: Cambridge University Press.

Legg, Catherine, and Christopher Hookway. 2020. Pragmatism. In *The stanford encyclopedia of philosophy* (Fall 2020 Edition), ed. Edward N. Zalta. https://plato.stanford.edu/archives/fall2020/entries/pragmatism/. Accessed 01 June 21.

Liszka, James. 2021a. *Charles peirce on ethics, esthetics and the normative sciences*. New York, NY: Routledge.

Liszka, James. 2021b. Pragmatism and the ethic of meliorism. *European Journal of Pragmatism and American Philosophy* 13.2. https://journals.openedition.org/ejpap/2442

More, Max. 2013. The philosophy of transhumanism. In *The transhumanist reader*, ed. M. More and N. Vita-More, 3–17. Chichester: Wiley-Blackwell.

Pearce, Trevor. 2020. *Pragmatism's evolution*. Chicago: University of Chicago Press.

Peirce, Charles S. *The Charles S. Peirce Papers*, Houghton Library, Harvard University. In-text citations to this collection begins with "R" and refer to the system of categorization in R.S. Robin, *The Annotated Catalogue of the Charles S. Peirce Papers*, and. R.S. Robin, *The Peirce Papers: A Supplementary Catalogue*.

Peirce, Charles S. 1931–1958. *Collected papers of Charles S. Peirce*, eds. Charles Hartshorne, Paul Weiss, and Arthur Burks. Cambridge, MA: Harvard University Press. In-text citations begin with "CP" followed by volume and paragraph numbers, and, when known, date of composition.

Peirce, Charles S. 1976. *The new elements of mathematics by Charles S. Peirce. Volume I Arithmetic, Volume II Algebra and Geometry, Volume 3.1 and 3.2 Mathematical Miscellanea, Volume IV Mathematical Philosophy*, ed. Carolyn Eisele. The Hague: Mouton Publishers. In-text citations begin with "NEM" followed by the volume number and the page number.

Peirce, Charles S. 1998. *The Essential Peirce*, vol. 2, eds. Nathan Houser, Christian Kloesel, and The Peirce Edition Project. Bloomington, Indiana: Indiana University Press. In-text citations begin with "EP" followed by the volume number and the page number.

Quine, W.V.O. 1981. The pragmatist's place in empiricism. In *Pragmatism: Its sources and prospects*, ed. R.J. Mulvaney and P.M. Zeltner, 21–39. Columbia, SC: University of South Carolina Press.

Raposa, Michael. 2018. Instinct and inquiry: A reconsideration of peirce's mature religious naturalism. In *Pragmatism and Naturalism: Scientific and Social Inquiry After Representationalism*, ed. M. Bagger, 27–43. New York Chichester, West Sussex: Columbia University Press.

Rorty, Richard. 1982. *Consequences of pragmatism*. Minneapolis: University of Minnesota Press.

Rorty, Richard. 1990. Science as solidarity. In *Objectivity, Relativism, and Truth: Philosophical Papers*, 35-45. Cambridge: Cambridge University Press.

Slater, M. 2015. *Pragmatism and the philosophy of religion*. Cambridge, UK: University of Cambridge Press.

Wilson, Aaron. 2016. *Peirce's empiricism: Its roots and its originality*. Lanham, MD: Lexington.

Wilson, Aaron. 2020. Interpretation, realism, and truth: Is Peirce's second grade of clearness independent of the third? *Transactions of the Charles S Peirce Society* 56 (3): 349–373.

Wilson, Aaron. 2018. Peirce's hypothesis of the final opinion. *European Journal of Pragmatism and American Philosophy* X(2). https://doi.org/10.4000/ejpap.1319

Aaron Bruce Wilson is Associate Professor of Philosophy at South Texas College, in McAllen, Texas. He is the author of Peirce's Empiricism: Its Roots and Its Originality (Lexington 2016) and of several articles on the philosophy of Charles S. Peirce. He grew up in New Hampshire, earned his Ph.D. in Philosophy from the University of Miami under professor Susan Haack in 2014, and in 2015 he won the Peirce Essay Prize from The Charles S. Peirce Society, in which he continues to be an active member. Aside from Peirce, he has research interests in Pragmatism, Early Modern Philosophy, Philosophy of Technology, Metaphysics, and Epistemology.

Chapter 3
Beyond Disintegration: Transhumanism and Enactivism

Marilyn Stendera

Abstract The enactive approach is becoming increasingly influential within the philosophy of cognition, to the extent that it is now one of the dominant models of embodied cognition—an umbrella term for a varied set of discourses sharing the view that our minds don't just happen to be 'in' bodies, but are enabled, shaped and (at least partly) constituted by the specifics of our physicality. This chapter will argue that the rise of enactivism is particularly relevant to transhumanist discourses, and vice versa, because their concerns intersect and conflict in vital ways. The discussion will use three core enactivist themes—organisational integrity, embodiment, and precarity—to draw out the kinds of tensions and intersections that enable enactivism and transhumanism to problematise one another. Enactivism defines life and cognition in terms of autonomy; that is, it posits that living systems generate and maintain themselves as porous yet bounded self-unities. This sets up a delicate balance—both for the enacting system and for enactivism itself—between the dual imperatives of adaptive self-creation and homeostasis. The system must change constantly in order to sustain itself, yet there is a limit to the system's flexibility. Beyond a certain point, change means disintegration, and disintegration means death. This balance itself resonates within transhumanist discourses, in the tension between the promise of radical self-transformation and the concern about taking this too far. These discourses, however, also challenge enactivism's potential to capture the full potential of the kinds of systems it describes. How do we determine the limits of morphological flexibility for cognisers as complex as ourselves? Are those limits fixed or malleable—and must integration always mean death, or can it facilitate redefinition?

3.1 Introduction

The enactive approach is becoming increasingly influential within the philosophy of cognition, to the extent that it is now one of the dominant models of embodied

M. Stendera (✉)
University of Texas Medical Branch, Galveston, US
e-mail: marilyn.stendera@deakin.edu.au

cognition—an umbrella term for a varied set of discourses sharing the view that our minds don't just happen to be 'in' bodies, but are enabled, shaped and (at least partly) constituted by the specifics of our physicality. This chapter will argue that the rise of enactivism is particularly relevant to transhumanist discourses because their concerns intersect and conflict in vital ways. The discussion will use three core enactivist themes—organisational integrity, embodiment, and precariousness—to draw out the kinds of tensions and intersections that enable enactivism and transhumanism to problematise one another.

Enactivism defines life and cognition in terms of autonomy; that is, it posits that living systems generate and maintain themselves as porous yet bounded self-unities. This sets up a delicate balance—both for the enacting system and for enactivism itself—between the dual imperatives of adaptive self-creation and homeostasis. The system must constantly change to sustain itself, yet there is a limit to the system's flexibility. Beyond a certain point, change means disintegration, and disintegration means death. This balance itself resonates within transhumanist discourses, in the tension between the promise of radical self-transformation and the concern about taking this too far. These discourses, however, also challenge enactivism's ability to capture the full potential of the kinds of systems it describes. How do we determine the limits of morphological flexibility for cognisers as complex as ourselves?[1] Are those limits fixed or malleable—and must disintegration always mean death, or can it facilitate redefinition?

According to enactivism, moreover, the specificities of a cogniser's embodiment matter. The system's physicality shapes the concerns it will pursue, the world it enacts for itself, and the means by which it does so. On the one hand, the enactive approach thus opens up another way of conceptualising why changing the parameters of our embodiment matters. Not only enhancements but any significant modifications to our bodies can change, enrich, enlarge, reduce, threaten our world, possibilities, and cognitive processes—which makes issues of regulation and access that much more poignant. On the other hand, the enactive approach amplifies the concern that we may no longer be who we are if we change our embodiment too radically; there may be limits to the circumstances under which we can cognise in recognisably human ways.

The enactive definition of cognition and life also means that both are characterised by an inherent precariousness. The system maintains its unity against the threat of disintegration. It must actively strive to maintain homeostasis because there is the continued possibility that it will fail, that external forces will disrupt its organisational unity. It is mortal by definition; for enactivism "life is precious because it is precarious" (Froese 2017). This forces us to ask how the kind of cogniser we are and the kind of world we enact would change if we were to change the limits of our

---

[1] A cogniser is a system that is capable of cognition. For enactive approaches, this means that it must be able to undertake the kind of sense-making outlined in Sect. 3.2, which requires autonomy and adaptivity (at least according to those who accept Di Paolo's work on the latter; see Di Paolo 2005). These terms will be defined later in the chapter. For now, it is worth nothing that the enactive model of cognition is particularly broad, embracing a vast range of different types of systems, arguing for what Thompson calls a "deep continuity of life and mind" (2007, p. 222).

precariousness, whether by radically decreasing it or even by seeking to transcend it altogether. If enactivism captures something true about what we are—if we are self-generating and self-sustaining systems—what happens if the processes involved in the latter change radically in scope? And what if precariousness is only redefined for some of us?

3.2   What Is Enactivism?

Enactive approaches are part of a broader set of discourses that, while heterogenous, share the view that cognition cannot be adequately captured in computational terms. According to classical computationalism, cognition consists in the manipulation of atomistic elements of symbolic systems according to syntactical rules, such that complex processes can be analysed into simpler constituents. The basic elements modified by these rules are representations as traditionally conceived, that is, context-independent in two ways.[2] Firstly, they codify context-independent information about properties, states of affairs, and so forth. Secondly, such representations are not taken to be significantly dependent upon or shaped by the cogniser's non-neural context. The cogniser's specific forms of embodiment, environment, or socialisation are taken to be (at most) 'quirks of the hardware' that are not essential to understanding the representation itself, nor contribute to the representation in a way that would prevent a cogniser with different specifications from working with the representation. Proponents of what are now often referred to as 4E views of cognition—encompassing not only enactive, but also embedded, embodied, and extended models of mind—reject this model and instead view cognition as a process that is shaped by, and can only be understood with reference to, the cogniser's particular non-neural bodily and environmental context.[3] In light of this, the four 'Es' tend to view cognition as primarily action-oriented, with an interest in explaining cognisers'

---

[2] I emphasise 'as traditionally conceived' here because there are other models of representation that do not include these characteristics; many of them have been proposed by proponents of 4E approaches. There is some controversy over whether any of these types of representation might be compatible with enactivism.

[3] Since all four 'Es' share an emphasis on the role of the non-neural body in cognition, it might seem strange that 'embodied' cognition is given an 'E' of its own, so to speak, or that enactivism is then also referred to as a type of embodied cognition at the beginning of the chapter. On the one hand, all of the Es do give the body a greater role than computational approaches to the mind, making all of the Es 'embodied' to some extent (for this reason, 'embodied cognition' and '4E cognition' are sometimes used interchangeably in the literature). On the other, each E treats and weights embodiment differently. Enactivism, for example, tends to assign a greater importance to the material specificities of particular types of bodies, while extended cognition is more closely aligned to functionalism. There are also models of embodied cognition that are neither extended nor enactive—hence the separate 'Es' here.

purposive and flexible responsiveness to salience in terms of their specific capacities, needs and ends.[4]

What, then, characterises the 'E' that is the focus of this chapter? Enactive approaches to cognition arose out of Humberto Maturana and Francisco Varela's work on defining life. They proposed that we can characterise living systems in terms of what they called autopoiesis. An autopoietic system continuously generates and specifies its own organisation through its operation as a system of production of its own components, and does this in an endless turnover of components under conditions of continuous perturbations and compensation of perturbations. (Maturana and Varela 1980, p. 79).

Such systems are autonomous, meaning that "they subordinate all changes to the maintenance of their own organisation" (p. 80) rather than to the achievement of an externally defined end. Their unity and identity are self-produced rather than being defined by an external observer or designer, and their complex responsiveness to changing circumstances cannot be reduced to a simple correspondence between inputs and outputs (pp. 80–81).

Autopoietic theory was initially targeted at the most basic living system, the cell. Recognising the broader value of its insights, however, core aspects of this approach were scaled up to allow their application to domains like cognition. The most significant step in this process was arguably taken by Varela himself, along with Evan Thompson and Eleanor Rosch, in their landmark 1991 work The Embodied Mind. This book wove together autopoietic theory with influences from the phenomenological tradition, cybernetics, developmental psychology and Buddhist philosophy to construct the framework for what is now known as enactive cognitive science. Three key aspects are worth highlighting for our present purposes. Firstly, the book shifted the focus to autonomy (of which autopoiesis is the most fundamental type).[5] Secondly, it argued for the ineluctable entanglement of perception and action.

Cognitive structures and processes emerge from recurrent sensorimotor patterns of perception and action. Sensorimotor coupling between organism and environment modulates, but does not determine, the formation of endogenous, dynamic patterns of neural activity, which in turn inform sensorimotor coupling (Thompson 2005, p. 407).

That is, the cogniser is always "structurally coupled" (Varela et al. 2016, p. 156) to its environment, and its cognition is characterised by multiple feedback loops that are enabled and shaped by its embodiment, by the particular sensorimotor capacities it has and the needs it is required to fulfil in order to maintain itself. Finally, in a point closely related to this, these cognitive processes do not reveal a predetermined world that merely impinges upon the cogniser from the outside. Instead, cognition is "a history of structural coupling that brings forth a world" (p. 209, my italics).

---

[4] My account of the opposition between computationalism and 4E approaches here draws primarily on Dreyfus (1972), Newen et al. (2018), and Thompson (2007).

[5] This point is still controversial within enactivist scholarship. Some couch their analyses primarily in terms of autopoiesis, while others emphasise autonomy (at least at the level of human cognition). This chapter will focus on autonomy mostly in order to circumvent these discussions.

Through its coupling with its environment, the cogniser enacts a world of significance defined in relation to its needs and ends; there is no cogniser without a world and no world without the cogniser. Only through the latter's particular capacities and projects do specific physical, chemical and biological aspects of the environment become nutrients or poisons, obstacles or tools, risks, opportunities, threats.

The three decades since the first publication of The Embodied Mind have seen a variety of developments to the enactive approach.[6] One that is especially significant to the intersection between it and transhumanist perspectives is Ezequiel Di Paolo's proposal that cognition also requires adaptivity, that is, the ability to respond to self-generated norms of flourishing (Di Paolo 2005). According to Di Paolo, a genuinely cognitive system must do more than produce and maintain itself; it must also be able to track whether it is doing better or worse at meeting its needs and staving off disintegration, and adjust accordingly. Another development worth noting here is the growth in the number and variety of analyses that draw on the enactive approach to some extent. Aspects of the enactive framework are being applied to the analysis of, among other things, educational design (e.g. Li et al. 2010), entrepreneurship (e.g. Fenwick 2010), nursing practices (e.g. Ousey and Gallagher 2007), neurodivergent experiences (e.g. De Jaegher 2020), musical performance (e.g. Høffding 2018), assistive technologies (e.g. Froese et al. 2012), narrative (e.g. Caracciolo 2014), art (e.g. Carvalho 2019) and film (e.g. Rhym 2018). While some encounters between transhumanist and enactive perspectives have already occurred, the latter's diversity of scope and influence means that these conversations are bound to proliferate. Given what enactive approaches say about the enabling conditions of our cognition—of our very being in the world—this dialogue is both urgent and likely to reveal productive tensions. The rest of this chapter will trace out three core aspects of enactivism that generate such points of intersection and conflict.

3.3 The Whole and Its Parts: Organisational Integrity

The first critical junction that I want to explore here is the enactive approach's emphasis on the maintenance of organisational integrity. An autonomous, adaptive system must navigate a delicate balance between two equally vital imperatives. On the one hand, it is an inherently dynamic system. In order to keep itself alive,

---

[6] A further development that has become especially relevant in the past decade is the 'splitting', for lack of a better word, into three main strands of enactive discourse: One, associated with figures like Thompson and Di Paolo, has continued the focus on the key themes of The Embodied Mind. (This is usually labelled 'autopoietic enactivism', although Di Paolo and Thompson point out that this is inaccurate due to the focus being on autonomy in general rather than just the basic autopoietic variety. See Di Paolo and Thompson 2014). A second approach deals almost exclusively with the structures of perception. The third and most recent type—'radical enactivism'—is mainly concerned with providing an account of what it calls 'basic minds', which involves extending the rejection of traditional representations to representations of all types as well as to content itself (Ward et al. (2017) provides more details about the relations and divergences between the three). This chapter will only engage with the first approach.

it must constantly track and respond to changes, not only in its environment and its own wellbeing, but also in the relation between them. Moreover, there will be other systems like it—ones with needs and aims and projects, forming not only potential threats or allies, but co-world builders with whom it can engage in "participatory sensemaking" (De Jaegher and Di Paolo 2007; De Jaegher 2019). On the other hand, such a system is defined by its need to maintain its unity and individuality, to persist as itself. It is a relational entity, and its boundaries are porous, yet they are real and essential nonetheless. If a change exceeds the system's ability to "compensate", in Maturana and Varela's terms, the result is "disintegration" (1980, p. 81). The rupture of boundaries, the loss of identity and individuality, lead to dissolution, to death. The enacting cogniser, then, must always change, yet never too much; both stasis and radical disruption are fatal. This tension recalls a familiar theme within debates about the benefits and risks of transhumanism. Many proponents of radical body modification and enhancement position these endeavours as an expression of autonomy, a continuation of the kinds of capacities that have positively shaped human development so far—the ability to adapt, the need to improve, the desire to thrive as well as survive (See e.g. Bostrom 2013; More 2013; Sandberg 2013). These claims, of course, face the well-known concerns about whether there is a point at which these transformations start to undermine something that defines us (Ross 2020). There are worries about drawing lines, about being able to recognise the transition from desirable to undesirable change, especially if the very processes that alter us also re-shape our views about what we are (and our ways of gauging how much change we are prepared to accept). In a sense, these debates enact on a large scale a question that, if enactivism is right, defines us—along with all other autonomous, adaptive systems, down to our very own cells: What is the right amount of change, the level that will let us survive without dissolving us?

This might make transhumanism and enactivism particularly congenial interlocutors, especially if each can learn from the other about different ways to address the question. Cary Wolfe's work on autopoietic theory and posthumanism is illustrative here.[7] Wolfe focuses on the distinction that Maturana and Varela draw between a system's organisation and its structure. The former refers to "those relations that must exist among the components of a system for it to be a member of a specific class" (Varela and Maturana, cited in Wolfe 1995, p. 52). The latter, meanwhile, is the "components and relations that actually constitute a particular unity" (p. 52). That is, a system's organisation cannot be altered without it losing its identity and dissolving, while its structure is more flexible and can undergo significant modifications. Indeed, it must do so; these are the kinds of changes that a system undergoes due to its coupling with its environment as well as its interactions with others like it. Autopoietic systems, in Wolfe's words, are "both open and closed" (p. 52) a way that

---

[7] Wolfe explicitly focuses on post-, rather than trans-, humanism. The distinction between them is, of course, controversial. I follow Wolfe (1995, 2010) and Harfield (2013) in viewing posthumanism as focusing more on a critique of humanism (especially in terms of anthropocentrism and the privileging of a certain model of rationality). However, I don't take this to be a hard and fast distinction, and follow Ross (2020) in thinking that these vast, disputed, heterogenous regions of discourse are close enough that insights about one can apply to the other.

he claims resonates deeply with the posthumanist critique of distinctions between the human and the non-human, nature, and culture, self and body and world. On the one hand, Wolfe argues, the way that such a system enacts its world—and the consequence that differences in organisation will lead to differences between such worlds—renders "the environment [in the sense we have been using world], and with it 'the body' [...] a virtual, multidimensional space" (2010, p. xxiii). On the other, what was previously a rigid, uncrossable ontological boundary between two sides of the distinction—between nature and culture, between the biological and the mechanical, and so on—is now made dynamic and, as it were, portable in the sense that the same formal mechanism may now be used to think, and link, across what were in the past discrete ontological domains (p. 206).

This means that for autopoietic theory, just as for posthumanism, "there can be no talk of purity" (p. xxv).

On this level, the potential conceptual sympathies between the two perspectives may go even further than Wolfe proposes here. One of the key consequences of autopoietic theory and the enactive approach founded upon it is that all cognitive systems—from the most basic to the most complex, whether organic or artificial— share the same fundamental structures: Autonomy and adaptivity. This means that enactivism aligns, not only with the posthumanist critique of boundaries and hier-archies but also with its rejection of anthropocentrism and concomitant affirmation of non-human importance (See Hartfield 2013). We are more complex than single-celled bacteria, but we share something fundamental with them—something more concretely defined than a mysterious essence of life. More than this, we are already machines: Living, autopoietic machines that are in turn comprised of concatenations of systems; we are, as in the title of one of Varela's essays, "a meshwork of selfless selves" (cited in Froese 2017, p. 38).

However, the very distinction that Wolfe focuses on—between organisation and structure—also constitutes a point of potential tension if we shift the lens from posthumanism to transhumanism. This is because differentiating between organisa-tional and structural integrity does not dissolve the concern about how much change a system can take; it just gives it a more precise target. The concept of organisational integrity asserts that there is such a limit; regardless of how structurally malleable and adaptive a cogniser may be, there are some types of transformations that will lead to disintegration. For someone interested in modifying and augmenting the body, the question then becomes how we can decide whether a particular change would be structural or organisational for the type of cogniser that we are. It seems that at least some technological enhancements of our physiological capabilities would be the former rather than the latter. The enactivist approach itself has been used to develop technologies allowing sensory substitution (e.g. the enactive torch, which provides haptic and auditory feedback to compensate for reduced vision—see Froese et al. 2012). More radical alterations to perception, however, might raise questions about whether the cogniser's world—enacted through its sensorimotor couplings—remains the same. These concerns would be amplified for technologies that go 'deeper', so to speak, and reach the heart of our self-producing, self-maintaining processes. Would certain types of gene therapy, for example, inaugurate organisational changes if they alter how the "selfless selves" (Varela 1991) that comprise us produce their components?[8] How much of our materiality can be replaced with radically different substances before a structural change becomes organisational? There is also the question of maintaining ourselves as auto—rather than allopoietic systems. Recall that, for autopoietic theory, the unity and individuality of a living system must be self-generated. Its ends and its boundaries must originate from itself, rather than being determined by the perspective of an external observer or designer. If this is the case, then we might wonder whether it is possible to compromise this—for example, that some types of implants or interfaces would mean that our boundaries are no longer self-originating; or that we might alter parts of ourselves to suit a specific purpose to the extent that we start to have externally-defined and designed ends. These questions, of course, do not only run one way. The enactivist, too, might wonder whether changes in the way that we relate to ourselves, and in the capabilities we have for transformation, should motivate a re-conceptualisation of organisational and structural terms. Perhaps we need to leave space within our models for a type of disintegration that leads to redefinition rather than annihilation—for example, by defining different levels of organisation change.

3.4 Bound(ed) Flesh: Embodiment

These concerns about classifying various modifications to ourselves as either structural or organisational also give us cause to look more closely at the enactive model of embodiment, which brings us to the second facet of enactivism that I want to explore here. As noted earlier, for enactivism, cognition is embodied in a radical way. The specificities of a cogniser's embodiment—its sensorimotor capacities, its needs, its specific means of motility and orientation—not only affect but enable cognition; that is, embodiment does not just have a contributory role, but a necessary, constitutive one.

This puts enactivism at odds with some of the more radical proposals under discussion in various transhumanist discourses, such as mind uploading. For one, the enactive approach denies the possibility of disembodied cognition—indeed, it makes this a conceptual impossibility—and therefore rejects any models of cognitive augmentation that see as their end goal the existence of a consciousness with no boundaries or sensorimotor feedback loops, the free-floating streams of virtual data familiar to us from science fiction versions of mind uploading. Of course, many contemporary models of the latter do not advocate for this, and instead suggest

---

[8] Varela's memorable descriptor "selfless selves" comes from the title of a 1991 chapter and refers to what Froese calls the "nesting" (2017, p. 38) way in which many small, basic autonomous systems can comprise larger, more complex ones (e.g. the way that cells form structures within our bodies, and these structures all add up to form us). Within these networks of overlapping processes and concerns, each autonomous unit is a 'self' (in the sense that it is a self-maintaining, self-preserving unity) and yet also 'selfless' (it does not possess a traditional sense of personal identity, and it is not isolated; its role within larger interlocking systems is important to making it what it is).

processes such as the gradual replacement of neurons by artificial neuron-like structures, the creation of a virtual body, or the transfer of neural processes to an artificial brain (or sufficiently brain-like artefact) connected to an organic or synthetic body (See Cappuccio 2017; Ross 2020 for discussions of these proposals). However, these do not resolve the tension. The enactive approach also challenges the general "neuro-centrism" (Cappuccio 2017) of approaches that downplay the role of the non-neural body in cognition, as if cognition could be 'unlocked' through the brain and everything else were just a secondary issue of finding the right matter to enable the transfer. As Cosmelli and Thompson argue, for the enactive model, the well-worn thought experiment of the 'brain in a vat' would simply not be plausible unless said 'vat' were a body like ours anyway, obviating the point of the exercise (2010). Even if we were able to secure a body much like ours for the uploaded mind, however, a deeper problem remains.

Cappuccio has argued that the real core of the conflict between embodied cognition and mind uploading lies, not in the issue of finding the right kind of material substrate for the mind, but in the assumption that the mind is the kind of thing that can be transferred between material substrates at all (2017). Enactivism—like other forms of embodied cognition—allows that minds can be instantiated in different types of materials; Maturana and Varela emphasised this right at the start (1980). However, Cappuccio argues, it must reject the claim that a mind instantiated in one type of material assemblage can be moved into another type of materiality while remaining qualitatively and numerically the same (2017). Mind uploading "posits criteria of continuity and identity of a mind that are extrinsic to its physical and functional constituents, and unrelated to the specific contextual integration of the mind–body-world system" (p. 438). For embodied models of cognition meanwhile, especially enactivism:

The patterns of these body-world interaction loops have a constitutive valence for the cognitive system but at the same time are merely relational in nature, i.e. situated, context-sensitive, non-exportable. Therefore, they are essentially irreplaceable in the unique way they are individuated in relation to neuronal and extra-cranial bodily interactions and to the beyond-the-skin world: that is why [embodied cognition] implies that the concrete instantiation of the mind in a contingent flow of material circumstances doesn't only define its functionality and phenomenology, but also its very conditions of ipseity and, therefore, the historically determined modes of its existence and persistence through time (p. 440).

Here, more than perhaps at any other point, we find a fundamental incompatibility between core enactivist claims and one type of transhumanist endeavour. Whether either side here is right will perhaps ultimately have to be determined in practice; if a version of mind uploading takes place, a host of discourses will need to re-evaluate critical aspects of their framework. This possibility in itself raises questions about how we would determine the success of such an event. How would we know it worked? Would we ask the uploaded one (presuming the result of the procedure is capable of responding)? This recalls the old concern about whether a mind deeply affected by artificial processes would be able to tell what it is. If the process destroyed the original system and created a new type of cogniser, the latter might nonetheless believe itself to be identical to the former. Ezequiel Di Paolo considers a similar issue in a recent paper, applying the enactive framework to the replicants of Ridley Scott's Bladerunner—some of whom famously are not aware of what they are (2020). Focusing on a point that the quote by Cappuccio also highlights—the historical determination of the mind—Di Paolo argues that the enactive approach ultimately speaks against the feasibility of implanted memories being sufficient to convince a replicant that they are human. This is because embodiment is historical. For Di Paolo, "these activities [of 'bodies in action'] do not only leave traces in (many) brains but practically everywhere. In my body and yours, in my surroundings, my shoes, my desk, my digital pursuits, and so on" (p. 22). This means that, while it is possible to create artificial bodies, "the idea that a full real bodily history can be faked" is "implausible" (p. 23). This suggests that it is not only embodiment but the history of embodied action, the temporal fabric of sensorimotor coupling, that is constitutive of cognition. Severing the link between a cogniser and its embodied history thus fundamentally alters, and possibly destroys, the former. Bringing this back to mind uploading, we can see here another way to support Cappuccio's claim about the non-exportability and irreplaceability of a mind's particular material instantiation. Moreover, we also find a hint of how the result of an upload might respond if the process did not work. Di Paolo finds it "hard to imagine" that Roy Batty would speak about his impending demise as he famously does in Blade Runner despite only having been alive for four years, and suggests that a mind lacking an embodied history may not even be able to engage in language, at least not in a way that we would understand (2020, p. 23). Perhaps this would also apply to the product of a mind upload. It might try to say that it is the same mind, but do so in a way that reveals it cannot be.

Of course, mind uploading is only one particularly drastic way of modifying bodily cognition. Would enactivists be similarly concerned about less radical changes? One response is that the enactive model of cognition at the very least gives us another way of understanding why altering our bodies matters. It decisively rejects the notion that such transformations are merely superficial or cosmetic. Recall that, for enactivism, cognition is "[a] history of structural coupling that brings forth a world" that works "[t]hrough a network consisting of multiple levels of interconnected, sensorimotor subnetworks" (Varela et al. 2016, p. 206). If cognition and the cogniser are inherently embodied in this way, then changing that embodiment not only changes who the cogniser is, what their projects might be and how they think, but also their world itself. This might seem like it would entail a negative response to body modification, yet it is important to remember that the cogniser's world is not static anyway. Just as the cognitive system can never stay still, so must its world remain dynamic, reshaped continuously in light of shifting significances generated by the cogniser's needs, ends and capacities, as well as its responsiveness to its physical and social environments. Structural change is, as we saw in the previous section, almost an imperative for the autonomous, adaptive system. Changing ourselves and our world is a defining feature of what we are, something we share with other cognisers; perhaps, then, those transhumanist voices who view body modification as an expression of deep-seated drives are onto something after all. On the other hand, this raises concerns for the impact upon our shared worlds. After all, our worlds are not ours alone; we enact them together.

Sensorimotor bodies, moreover, are enacted together. [...] There are in social encounters situations where the sensemaking of a participant is literally modulated or enabled by the activity of others, and in some cases, sensemaking is constituted jointly in co-authored social acts (Di Paolo 2020, p. 17).

In altering my embodiment, then, I am not only reshaping my world, but also ours—and the ability to make it ours. At what point, then, do changes to the bodies of some disrupt their ability to generate and navigate significance in concert with others? This adds another layer of urgency to questions about equity of access to augmentations. The concern that only some will be able to utilise such advances, entrenching existing axes of disadvantage and potentially creating new ones, is a familiar trope in transhumanist debates (Ross 2020; Sandberg 2013). Enaction opens up a further way to conceptualise what is at stake—namely, our ability to participate in shared world-building. Of course, we have always shared and made worlds with cognisers of different embodiments, so it seems that there is a certain amount of flexibility to 'participatory sensemaking'. The question then seems to be, again, one of finding a way to draw a line, of asking when our worlds are at risk of becoming irreconcilable.

3.5 Life as Perpetual Struggle: Precariousness

The final aspect of enactivism that I want to place in dialogue with transhumanist concerns is one that is already suggested by the idea of an inherently embodied system striving to preserve its organisational integrity. That is, the enacting cogniser is characterised by precariousness in a way that, I want to suggest here, both challenges and is challenged by transhumanist attitudes to the limitations of the human condition (See also Di Paolo 2020; Froese 2017). As noted earlier, the autopoietic, adaptive system is by nature dynamic, constantly adjusting in response to shifting relationships between environmental circumstances and its needs, capacities, and projects. These relationships, however, cannot be finished or perfected; there will always be gaps between what the system needs and what its environment supplies, between risks and rewards; even if circumstances are favourable, they can always change. Even the processes through which the system produces and maintains itself come with an inherent risk; complex cognisers especially need to keep their own components in check, lest they become, for example, cancerous threats to the whole. This perpetual threat of disintegration, however, is more than a constraint. After all, the very project of self-maintenance only makes sense if it is possible for that process to fail; self-individuation requires something against which and in the face of which the system must unify itself, bound itself, keep itself going. This is a life that defines itself through the possibility of its own end. To that extent, we might almost say that enaction is founded upon a perpetual negation.

If the enacting cogniser is characterised by precariousness to this extent, then it becomes difficult to reconcile this model of cognition with transhumanist endeavours aimed at radically reducing or even transcending such limitations. At first glance, this may seem like just another version of the familiar concern that struggle is what makes life worth living, that the inevitability of death somehow gives human life value (Ross 2020). While this rings true to some extent, there is nonetheless more to the enactivist angle here. Tom Froese, for example, argues that the precariousness of the enactive cogniser is what enables it to have any concerns or projects in the first place—to enact significance, to bring forth a meaningful world (2017). For enactivism, "to live is to always be concerned with something, most fundamentally with the continuation of one's individual manner of living" (p. 24). It is the imperative to survive in the face of potential annihilation that lets the system generate the most basic meanings—nutrition, lack, threat, and so forth. The struggle with precariousness generates the first and ultimate endogenous ends; it is the reason that anything at all can matter to the system. According to Froese, this means that taking seriously the biologically embodied mind cannot avoid bringing us face to face with the inevitability of our own finitude, which conflicts with the transhumanist goal of defeating death by engineering our bodies to stay forever young (p. 47).

It is important to clarify here, of course, that the transhumanist perspectives to which Froese is referring here are not advocating for immortality as such. Even Aubrey de Grey frames his goals in terms of 'amortality', not only to avoid the conceptual baggage that the more familiar term brings with it, but also to acknowledge that the augmented individual could still die (Ross 2020). The issue, then, is not so much one of escaping precariousness as of radically modifying its parameters. Indeed, one might say that a human cogniser who seeks to extend their life is only expressing the fundamental self-maintaining striving that characterises all autonomous, adaptive systems; perhaps amortality is taking enaction to its limits. One concern here might be that this is a self-undermining endeavour. If precariousness is an enabling condition of cognition, then the cogniser that successfully eliminates it thereby brings to an end its way of being. This is a recurring theme within a vast range of discourses—that we are characterised by a lack whose overcoming would be our destruction, that "nothing finished can live" (Jaspers 1970, p. 200). Di Paolo articulates this in terms of the incompatibility of perfect self-production or individuation with life.

In neither case, maximal self-production or maximal self-distinction, do we have a living system. The dialectical resolution of this tension is the regulated deferral of openings and closings to environmental influences that keep the system viable. Such regulation with respect to viability conditions is what we have called sensemaking (2020, p. 16).

The transhumanist could still respond here that even significant extensions to one's lifespan would be reducing and reformulating, rather than eliminating, this vital precariousness. As Nick Bostrom writes, the "posthuman could be vulnerable, dependent, and limited" (2013, p. 48). However, at least two concerns would continue to generate tensions between this perspective and the enactive approach. On the one hand, we face another version of the point about irreconcilable worlds that was raised at the end of the previous section. Precariousness enables, shapes, and constrains the enaction of a world of significance, which means that changing the former changes the latter. As noted before, we make worlds with cognisers whose embodiment differs from ours; surely, this also applies to the specifics of precariousness. However, as with embodiment, we might wonder whether there is a point at which modifications to our precariousness interfere with our ability to make and navigate meaning together, where our projects become more difficult to weave together with the interests of those whose lives are much more or less precarious than our own. On the other hand, there may be what Froese calls a "mismatch at the conceptual level: Transhumanism views mortality as a burden to be removed or at least as something to be postponed indefinitely by scientific progress, rather than as constitutive of a meaningful way of life" (2017, p. 47). We can extend this beyond mortality to other limitations, which transhumanist discourses tend to cash out as something to be overcome, even if they cannot be left behind altogether (e.g. More, 2013; Bostrom, 2013). Striving beyond them is an imperative, something that we should at least try to do. For the enactive approach, meanwhile, limitations are often also enabling conditions. In saying this, it is important to avoid relegating enactivism to what More calls "apologism—the view that it is wrong for humans to attempt to alter the conditions of life for the better" (2013, p. 14). As detailed in the first section, continuous change, as well as the aim to do well and better according to its own standards of flourishing, define the autonomous, adaptive system; the cogniser must engage in structural modifications in pursuit of these. Even in light of this clarification, we might still worry that the emphasis on precariousness could lead to the veneration of suffering and hardship. However, we must not confuse the basic limitations of cognition with particular forms that they might take at the 'macro' level. The former do not necessitate or legitimise the latter. Moreover, the claim is not that it is 'good' or 'right' that cognisers must maintain a fragile unity in the face of internal and external threats, that they must operate by means of imperfect feedback looks to resist disintegration. Rather, these are simply the necessary conditions for us being in any way at all. Limitless cognition is an oxymoron.

3.6 Concluding Remarks

Does all of this mean that enactive approaches to cognition are by nature bioconservative? One point worth considering is that enactivism's resistance to some transhumanist endeavours is grounded in very different concerns to those of more familiar critiques. That is, the enactive perspective outlined here does not proceed from the assumption that humans are fundamentally different to all other entities, nor does it argue for some mysterious human essence or telos that must be preserved (See Harfield 2013; More 2013; Ross 2020). Instead, it takes the opposite approach. The tension with transhumanist imperatives is not generated by what sets us apart, but by what we share with all cognisers—autonomy and adaptivity, the need to preserve organisational integrity while negotiating structural changes, the coupling of body and world, precariousness. This arguably makes the prospect of a sustained dialogue between the perspectives particularly promising. On the one hand, the enactivist challenge is couched in terms that are themselves at least minimally congenial to transhumanist (and posthumanist) discourses. Both perspectives suggest that we are dynamic rather than static creatures, that our bodies matter and that distinctions between the human and non-human are neither straightforward nor rigid. On the other hand, transhumanist projects are well-suited to function as test cases for enactive models of cognition. If the former can achieve something that the latter claim should not be possible, then the autonomy and adaptivity, organisational integrity, embodiment, and precariousness may need to be radically reconceptualised.

References

Bostrom, Nick. 2013. Why I want to be a posthuman when I grow up. In *The transhumanist reader*, ed. M. More and N. Vita-More, 28–53. Chichester: Wiley-Blackwell.

Cappuccio, Massimiliano Lorenzo. 2017. Mind upload: The ultimate challenge to the embodied mind theory. *Phenomenology and the Cognitive Sciences* 16: 425–448. https://doi.org/10.1007/s11097-016-9464-0.

Caracciolo, Marco. 2014. *The experientiality of narrative: An enactivist approach*. Boston: De Gruyter.

Carvalho, John M. 2019. *Thinking with images: An enactivist aesthetics*. New York: Routledge.

Cosmelli, Diego, and Evan Thomson. 2010. Embodiment or envatment? Reflections on the bodily basis of consciousness. In *Enaction: Toward a new paradigm for cognitive science*, ed. J. Stewart, O. Gapenne, and E.A. Di Paolo, 361–385. Cambridge MA: MIT Press.

De Jaegher, Hanne. 2019. Loving and knowing: Reflections for an engaged epistemology. *Phenomenology and the Cognitive Sciences*.https://doi.org/10.1007/s11097-019-09634-5

De Jaegher, Hanne. 2020. Seeing and inviting participation in autistic interactions. *Transcultural Psychology*, forthcoming.

De Jaegher, Hanne, and Ezequiel Di Paolo. 2007. Participatory sensemaking. *Phenomenology and the Cognitive Sciences* 6: 485–507. https://doi.org/10.1007/s11097-007-9076-9.

Di Paolo, Ezequiel. 2005. Autopoiesis, adaptivity, teleology, agency. *Phenomenology and the Cognitive Sciences* 4: 429–452. https://doi.org/10.1007/s11097-005-9002-y.

Di Paolo, Ezequiel, and Evan Thompson. 2014. The Enactive Approach. In *The Routledge Handbook of Embodied Cognition*, ed. Lawrence Shapiro, pp. 68–78. New York: Routledge.

Di Paolo, Ezequiel. 2020. Enactive becoming. *Phenomenology and the Cognitive Sciences*. https://doi.org/10.1007/s11097-019-09654-1

Dreyfus, Hubert. 1972. *What computers can't do: A critique of artificial reason*. New York: Harper & Rowe.

Fenwick, Tara J. 2010. Work knowing "On the Fly": Enterprise cultures and co-emergent epistemology. *Studies in Continuing Education* 23: 243–259. https://doi.org/10.1080/01580370120101993.

Froese, Tom. 2017. Life is precious because it is precarious: Individuality, mortality and the problem of meaning. *Studies in Applied Philosophy, Epistemology and Rational Ethics* 28: 33–50. https://doi.org/10.1007/978-3-319-43784-2_3.

Froese, T., M. McGann, W. Bigge, A. Spiers, and A.K. Seth. 2012. The enactive torch: A new tool for the science of perception. *IEEE Transactions on Haptics* 5: 365–375. https://doi.org/10.1109/TOH.2011.57.

Harfield, Timothy D. 2013. Exposing humanism: Prudence, ingenium, and the politics of the posthuman. *Journal of Historical Sociology* 26: 264–288. https://doi.org/10.1111/johs.12001.

Høffding, Simon. 2018. *A phenomenology of musical absorption*. Cham: Springer.

Jaspers, Karl. 1970. *Philosophy*, vol. 2, trans. E. B. Ashton. Chicago: University of Chicago Press.

Li, Qing, Bruce Clark, and Ian Winchester. 2010. Instructional design and technology grounded in enactivism: A paradigm shift? *British Journal of Educational Technology* 41: 403–419. https://doi.org/10.1111/j.1467-8535.2009.00954.x.

Maturana, Humberto, and Francisco J. Varela. 1980. *Autopoiesis and cognition: The realisation of the living*. London: D Reidel.

More, Max. 2013. The philosophy of transhumanism. In *The transhumanist reader*, ed. M. More and N. Vita-More, 3–17. Chichester: Wiley-Blackwell.

Newen, Albert, Shaun Gallagher, and Leon De Bruin. 2018. 4E cognition: Historical roots, key concepts and central issues. In *The Oxford handbook of 4E cognition*, ed. A. Newen, S. Gallagher, and L. De Bruin. Oxford: Oxford University Press. https://doi.org/10.1093/oxfordhb/9780198735410.013.1.

Ousey, Karen, and Peter Gallagher. 2007. The theory-practice relationship in nursing: A debate. *Nurse Education in Practice* 7: 199–205. https://doi.org/10.1016/j.nepr.2007.02.001.

Rhym, John. 2018. Historicizing perception: Film theory, neuroscience, and the philosophy of mind. *Discourse* 40: 83–109.

Ross, Benjamin. 2020. *The philosophy of transhumanism: A critical analysis*. Bingley: Emerald Publishing.

Sanberg, Anders. 2013. Morphological freedom—Why we not just want it, but need it. In *The transhumanist reader*, ed. M. More and N. Vita-More, 56–64. Chichester: Wiley-Blackwell.

Thompson, Evan. 2005. Sensorimotor subjectivity and the enactive approach to experience. *Phenomenology and the Cognitive Sciences* 4: 407–427. https://doi.org/10.1007/s11097-005-9003-x.

Thompson, Evan. 2007. *Mind in life: Biology, phenomenology and the sciences of the mind*. Cambridge MA: Belknap Press.

Varela, Francisco J. 1991. Organism: A meshwork of selfless selves. In *Organism and the origins of self*, ed. A.I. Tauber, 79–107. Dordrecht: Kluwer.

Varela, Francisco J., Evan Thompson, and Rosch Eleanor. 2016. *The embodied mind*. Revised. Cambridge MA: MIT Press.

Ward, Dave, David Silverman, and Mario Villalobos. 2017. Introduction: The varieties of enactivism. *Topoi* 36: 365–375. https://doi.org/10.1007/s11245-017-9484-6.

Wolfe, Cary. 1995. In search of post-humanist theory: The second-order cybernetics of Maturana and Varela. *Cultural Critique* 30: 33–70.

Wolfe, Cary. 2010. *What Is posthumanism?* Minneapolis: University of Minnesota Press.

Dr. Marilyn Stendera is a Lecturer in Philosophy at Deakin University. She received her Ph.D. (on the intersection between Heideggerian phenomenology and 4E cognitive science) from the University of Melbourne. Her current work focuses primarily on different conceptualizations of temporality, especially within the phenomenological tradition, and the dialogues between phenomenology and philosophies of cognition, mind, and science.

Chapter 4
Overcoming (Our) Nature: Transhumanism and the Redefinition of Human Being's Essence

Georgios Karakasis

Abstract The aim of this chapter is to shed a critical light on transhumanism's philosophy as regards its positions on nature, human nature, technology, and death. Founding my argumentation on Martin Heidegger's later thought I will show that transhumanism aims to radically change human nature, to create a new human being, in whom technology and the organic merge so as to finally overcome death, thus fulfilling, to the highest degree, modern human being's "will to life".

4.1 Introduction

The aim of this chapter is to examine transhumanism from a Heideggerian point of view, placing an emphasis on three main issues: (a) the essence of the human being as *deinon* and the former's constant struggle against *phusis* (nature); (b) the relation between the human being and "Positionality (Gestell)", as technology's essence; as well as (c) the necessity of our own most possibility of dying as a key element for the understanding of the meaning of our being in this world.

Taking as my point of departure Heidegger's *Introduction to Metaphysics*,[1] I will try to show that the human being, from the very first moment of his emergence in the world as being, is acting violently against *phusis*, against nature, feeling limited by it, and he desires to uncover all of nature's secrets, namely Being itself. Through transhumanism, what in the past was considered impossible, namely changing nature and our own human nature, becomes now an option since through the technological progress and the technical knowledge we have obtained we are now capable of redefining our own essence through the use of technology; in other words, the human being, having transhumanism as the spearhead of his assault, is openly claiming the crown of creator for himself.

---

[1] For a thorough analysis of this dense work see: Polt and Fried (2001).

G. Karakasis (✉)
University of the Basque Country, Donostia-San Sebastian, Spain
e-mail: geokarakasis7@yahoo.com

Next, presenting Heidegger's interpretation of technology's essence as "Position-ality", I will show that transhumanism is a decision made by the human being to offer himself and his essence to technology, with the higher goal of further enhancing his life, thus limiting the possibility of any possible apparition of physical restraints as a consequence of the aging or decaying of his bodily existence. Transhumanism becomes, thus, a decision made by human beings to start existing in another way, overcoming the "obstacle" of our bodies by merging the organic with the artificial for the sake of our "will to life".

Finally, I will examine the relation between transhumanism and death. It could be said that the "arch-enemy" of transhumanism is death itself since it puts an end to our lives and openly challenges the power of the human being. Transhumanism casts into doubt death's significance in our lives, deprives it of its ontological importance, and treats it solely as an obstacle to be overcome in our quest for eternal life. Basing my arguments on Heidegger I will show that death's existence in our lives, and our own mortality, are both elements of crucial importance since it is only through these that life has meaning, significance, and purpose, protecting us from the "eternal recurrence of now", of the deathless transhumanist vision. Death becomes the "refuge of Being", protecting Being from being limited to an objectified presence of ready-at-hand and disposable beings.

4.2 Yet Nothing Uncannier Than Man: Human Being as Deinon and Phusis as the Overwhelming Sway

"Πολλὰ τὰ δεινὰ κοὐδὲν ἀνθρώπου δεινότερον πέλει/Manifold is the uncanny, yet nothing uncannier than man bestirs itself (Heidegger 2014, p. 163)". This short yet dense passage comes from the first Choral Ode (lines 332–333) of Sophocles' work, *Antigone* and it is of paramount importance for Heidegger's understanding both of human being and his relation to phusis (φύσις)—nature in Greek—and the beings surrounding us. Heidegger uses the word uncanny, unheimlich in German, to translate the Greek word deinon (δεινόν). Following Heidegger, deinon is:

> The *deinon* is the terrible in the sense of the overwhelming sway, which induces panicked fear, true anxiety, as well as collected, inwardly reverberating, reticent awe. The violent, the overwhelming is the essential character of the sway itself. (Heidegger 2014, p. 166)

Understanding what deinon is demands an understanding of what the "over-whelming sway" stands for in Heidegger's thought. This overwhelming sway, in front of which fear, awe, and true anxiety are provoked, is the emergence of Being as phusis, in the inceptual thought of the ancient Greeks. Phusis, one of the key ideas throughout Heidegger's entire thought, is clearly distinguished from the beings that surround us. Phusis is not a sum of beings, nor something merely constant around us, no matter its immensity. Quoting Heidegger:

> *phusis* means the emergent self-upraising, the self-unfolding that abides in itself. In this sway, rest and movement are closed and opened up from an originary unity. This sway is the

overwhelming coming-to-presence that has not yet been conquered in thinking, and within which that which comes to presence essentially unfolds as beings. But this sway first steps forth from concealment—that is, in Greek, *alētheia* (unconcealment) happens—insofar as the sway struggles itself forth as a world. Through world, beings first come into being. (ibid., p. 67)

Phusis as emergence can be experienced everywhere: for example, in processes in the heavens (the rising of the sun), in the surging of the sea, in the growth of plants, in the coming forth of animals and human beings from the womb. But *phusis*, the emerging sway, is not synonymous with these processes, which we still today count as part of "nature." This emerging and standing-out-in-itself-from itself may not be taken as just one process among others that we observe in beings. *Phusis* is Being itself, by virtue of which beings first become and remain observable. (ibid., p. 16)

Thus, Being is phusis, and phusis is the overwhelming sway, the overwhelming coming to presence of beings, human beings included. We as human beings are brought forth by phusis through this violent sway and this is the only way we can stand forth in presence: as phusis and through phusis. We appear and we can see, grasp, and mentally represent other beings because we are all inside this overwhelming sway. Still, human beings have a special relation with the other beings, due to our very specific relation with Being. We human beings are the only beings whose Being is an issue for them; we care about our being and Being, and we also try to understand what Being is and find meaning in every action and relation in this world. Knowing that we *are*, we search everywhere for clues that may help us define who we really are, how and why we are, as well as what makes us who we are, namely Being. Heidegger comments concerning this unique relation we have with Being as follows:

Furthermore, if appearing belongs to Being as *phusis*, then the human, as a being, must belong to this appearing. And since Being-human amid beings as a whole manifestly constitutes a distinctive way of Being, the distinctiveness of Being-human grows from its distinctive way of belonging to Being as the appearing that holds sway. (ibid., p. 155)

Still, even though as human beings we have this privileged relation with Being, our interaction with it is not one characterized by passivity, recognition, peace, and well-being. The human being as deinon, according to Heidegger, is violence-doer, and this is the only way he can be inside the overwhelming sway of phusis. We are confrontational beings and our struggle to comprehend and apprehend our being is a history of strife, failures, and defeats. The human being releases all of his violence-doing because he wants to reveal Being in this world. This revelation of Being takes place through the human being's *techne*. Quoting Heidegger:

Because art, in a distinctive sense, brings Being to stand and to manifestation in the work as a being, art may be regarded as the ability to set to work, pure and simple, as *techne*. Setting-to-work is putting Being to work in beings, a putting-to-work that opens up (…) Thus *techne* characterizes the *deinon*, the violence-doing, in its decisive basic trait; for to do violence is to need to use violence against the over-whelming: the knowing struggle to set Being, which was formerly closed off, into what appears as beings. (ibid., p. 178)

Having seen all the above we are now getting closer to understanding why the human being is defined as deinon. The human being is part of the overwhelming sway of phusis—thus belonging to Being—but his relation with beings and Being can only be a violent one; the human being as deinon is a violence-doer because through the means of techne, deinon tries to uncover Being, bringing it into steady appearance and constancy in beings. Doing so, human beings are constantly forming the world since the world appears through beings and the meaning of the world is based on the relations developed (or not) inside this world. As human beings we are constantly struggling, and strife is our way of being due to the fact that our relation with Being is a conflictive one, a relation based on our violent acting against Being, even though we are part of its overwhelming sway. We are in phusis but that does not prevent us from endlessly challenging it, bringing it to its limits, demanding its total unconcealment in front of our eyes and for the sake of them. The human being as deinon, as the uncanny, can now be better understood in the following way, following Heidegger:

> Beings as a whole, as the sway, are the overwhelming, *deinon* in the first sense. But humanity is *deinon*, first, inasmuch as it remains exposed to this overwhelming sway, because it essentially belongs to Being. However, humanity is also *deinon* because it is violence doing in the sense we have indicated. [It gathers what holds sway and lets it enter into an openness.] (ibid., p. 167)

The definition of the human being given by Heidegger is likely to sound unfamiliar, uncanny, and/or even pessimistic. We are human beings because we are violent and subject to strife. The world for the human being is a miracle, a beautiful appearance and a perfect order brought forth by phusis; but when the human being decides to essentialize his essence as deinon then the world becomes a battlefield and there is a fierce assault against the origin, against the overwhelming sway of phusis. There are no limits, no boundaries during this struggle because the "sacred", phusis, and Being are put into question; they are challenged to appear in front of the violence-doer. There can be only one result, though: the total defeat of the human being by the more powerful overwhelming sway. Heidegger comments:

> The essence of Being-human opens itself up to us only when it is understood on the basis of this urgency that is necessitated by Being itself. Historical humanity's Being-here means: Being-posited as the breach into which the excessive violence of Being breaks in its appearing, so that this breach itself shatters against Being. (ibid., p. 181)

And, depicted in an even more explicit way:

> Dasein is the constant urgency of defeat and of the renewed resurgence of the act of violence against Being, in such a way that the almighty sway of Being violates Dasein (in the literal sense), makes Dasein into the site of its appearing, envelops and pervades Dasein in its sway, and thereby holds it within Being. (ibid., p. 198)

The world, thus, may be unfolded in the eyes of the human being; nevertheless, this unconcealment is not to be considered as a retreat by the world when faced with the human being. Quite the contrary, the human being is forced to bend his will when confronted with the overwhelming sway of phusis and Being. Being will appear in beings but not as a result of deinon's successful violence-doing, but as its total and unconditional surrender to a sway that far exceeds human beings' own power. Human beings will always be the violence-doers against phusis, we will always strive to make it reveal its secrets to us, but phusis will always shatter our violence-doing through its own violence: we will never bend Being and its meaning to our will since Being will always answer our challenge, shattering our inquisitive violence-doing and making us the site of its manifestation. In a nutshell, phusis' manifestation is always granted, never conquered.

4.3 Fixing the Flawed: Transhumanism's Assault Against Phusis

The reason why I have dedicated this first part to Heidegger's interpretation of the human being in phusis lies in the fact that the movement/philosophy of transhumanism[2] sets forth a radically different approach concerning the relations among the human being, Being and phusis. The philosopher and defender of transhumanism Max More, in an open letter to nature herself titled "A Letter to Mother Nature", says the following:

> Mother Nature, truly we are grateful for what you have made us. No doubt you did the best you could. However, with all due respect, we must say that you have in many ways done a poor job with the human constitution. You have made us vulnerable to disease and damage. You compel us to age and die – just as we're beginning to attain wisdom. You were miserly in the extent to which you gave us awareness of our somatic, cognitive, and emotional processes. You held out on us by giving the sharpest senses to other animals. You made us functional only under narrow environmental conditions. You gave us limited memory, poor impulse control, and tribalistic, xenophobic urges. And, you forgot to give us the operating manual for ourselves! (More 2013, p. 449)

The above text/confession concerning the role of phusis in human beings' life— even though we cannot say that it represents all of the proponents of the transhumanist philosophy—sheds a new light on how phusis is to be conceived in this philosophical and technological movement. Nature, namely phusis, has made a lot of mistakes when it comes to the formation of the human being, and seems to have a cruel sense of humor: our death arrives just when we start to become wise. The human being depicted in this letter probably has nothing to do with the Sophoclean and Heideggerian human being as deinon who dares to release all the violence kept inside to uncover Being in beings. On the contrary, the human being we see here is malfunctioning, afraid, and in constant need of abilities that he will never have if he remains dependent on nature. This reminds us of the Platonic Dialogue *Protagoras*, in which Prometheus tries to make up for the errors of Epimetheus, who molded human beings that lacked the most basic skills and abilities to survive in a world

---

[2] Nick Bostrom, co-founder of the World Transhumanist Association, defined Transhumanism in the following way: "The intellectual and cultural movement that affirms the possibility and desirability of fundamentally improving the human condition through applied reason, especially by developing and making widely available technologies to eliminate aging and to greatly enhance human intellectual, physical, and psychological capacities (Bostrom 2003)".

full of dangers, providing the human species with τέχναι (technai), the technics and skills needed to protect humanity from extinction.

Reading the above letter in this way, we see that transhumanism arises as a new Prometheus, offering its help to make up for the errors committed in regard to the human being. Nonetheless, there is one important difference in this case: the errors are made by phusis itself; hence, our creator, phusis, has made mistakes that we, human beings, as having emerged from phusis, are now obliged to undo and correct. Transhumanism, thus, has come to claim for itself phusis' "operating manual", paving the way for a new creation, freed from the mistakes of the past—mistakes that are not of a human nature but that require human intervention to be overcome and erased. The omnipotent subject is arising again in transhumanism, but this time it does not demand priority over the rest of creation; rather, it claims the role of creator for itself. What is unfolded here is the struggle between the human "will to life" and the limits set by phusis to halt the expansion of our lives. In this letter we can trace the Nietzschean "will to power" in its purest form as "will to life"—a life that is no longer limited to its self-preservation but now demands its unconditioned expansion. Quoting Heidegger:

> Nietzsche does not see the essence of life in "self-preservation" ("struggle for existence") as do the biology and the doctrine of life of his time influenced by Darwin, but rather in a self-transcending enhancement. As a condition of life, value must therefore be thought as that which supports, furthers, and awakens the enhancement of life. (Heidegger 1991, pp. 15–16)

The way Nietzsche conceived this life-related enhancement is also clearly depicted in the following passage:

> According to Nietzsche, self-preservation and self-enhancement belong together as fundamental features of life. Each self-preservation of life serves the growth of strength and power, because a life that is purely focused on self-preservation would already imply its demise according to Nietzsche. Therefore, the self-preservation is never the "goal" or "destination" of life as such, but always and only as a means for the goal of self-enhancement. (Blok 2017, p. 24)

In transhumanism, as well, life, together with the ways of enhancing life, seems to be the absolute value and the ultimate goal. Self-preservation is the first step leading to the greater breakthrough of humanity, ousting phusis from its place in order to guarantee the continuity of its own expansion. According to Nick Bostrom:

> Transhumanists view human nature as a work-in-progress, a half-baked beginning that we can learn to remold in desirable ways. Current humanity need not be the endpoint of evolution. Transhumanists hope that by responsible use of science, technology, and other rational means we shall eventually manage to become post-human, beings with vastly greater capacities than present human beings have. (Bostrom 2001)

What phusis started—and could not finish, at least successfully—transhumanism seeks to bring to another level and complete. Humanity is no longer restricted to its defensive role of protecting itself against phusis. The time is ripe for a full offensive against phusis—an offensive that will be made possible by the technological enhancement of our life. These enhancements vary, since they may include "radical extension of human health-span, eradication of disease, elimination of unnecessary suffering, and augmentation of human intellectual, physical, and emotional capacities. (ibid.)" Furthermore, according to Julian Savulescu "Enhancement is no luxury. In so far as it promotes well-being, it is the very essence of what is necessary for a good human life (Savulescu 2005, p. 38)".

So far, quoting different representatives of transhumanism, we have witnessed: the appearance of a human being that is only functional "under narrow environmental conditions"; a human essence that is still "half-baked" and needs to be completed; the crucial importance of enhancement in order to finally possess what is necessary "for a good human life"—probably in comparison to the less good life that we, non-enhanced people, currently live; and, finally, the eventual substitution of phusis by the technologically enhanced transhuman.[3] Technology, thus, would reach the highest level of its development and progress, having managed to merge with life and with the human being's essence. Technology would no longer be a tool, but, quite the opposite, a new way of living and being. Transhumanism is the apogee of the omnipotent "Cartesian maître et possesseur de la nature (Blok 2014, p. 309)" that is no longer limited to understanding the laws of nature, but can now rewrite them.

At this point the objection could be raised that we cannot know whether transhumanism will further damage or heal the wounds inflicted on phusis by the human being. It could also be the case that by "fixing" human nature, our attitude may also change toward phusis, becoming more respectful and ecological in our actions and thinking. Still, this does not change the ontological reality on which transhumanism is founded: namely, that we, human beings, are a priori in need of enhancement, which thus proves the inability of phusis to offer us what we deserve. It seems that, from the transhumanist perspective, actual humanity is born flawed, and in dire need of "repairing". The transhuman stands against the actual human being as the constant reminder that we are born physically defective and that our defects are our "natural" way of being in this world. Quoting Max More:

---

[3] The concept transhuman can be better understood when seen in comparison with the concept posthuman. Following Robert Ranisch and Stefan Lorenz Sorgner: "The link between the human and the posthuman is the transhuman, an abbreviation for a transitional human, to which transhumanism owes its name. In this regard, transhumanism can be understood as a *transhuman*-ism. By the same token, transhumanism, according to its self-understanding, is a contemporary renewal of humanism. It embraces and eventually amplifies central aspects of secular and Enlightenment humanist thought, such as belief in reason, individualism, science, progress, as well as self-perfection or cultivation (Ranisch and Lorenz Sorgner 2014, p. 8). Furthermore, according to Keith Ansell Pearson: "The transhuman condition is not about the transcendence of the human being, but concerns its nonteleological becoming in an immanent process of 'anthropological deregulation'. When Nietzsche asks his 'great' question, what may still become of man?, he is speaking of a future that does not cancel or abort the human, but one which is necessarily bound up with the inhuman and the transhuman (Ansell Pearson 1997, p. 163). Finally, as Philbeck (2014) puts it: The "posthuman" in posthumanism, on the other hand, refers to a state of being that is beyond our understanding from a humanist philosophical paradigm. It delineates an entity that defines and understands itself differently than through the contemporary notion of 'human' because of technology's impact on basic human characteristics. (p. 175).

> True transhumanism doesn't find the biological human body disgusting or frightening. It does find it to be a marvelous yet flawed piece of engineering, as expressed in Primo Posthuman. It could hardly be otherwise, given that it was designed by a blind watchmaker, as Richard Dawkins put it. True transhumanism does seek to enable each of us to alter and improve (by our own standards) the human body. (More 2013b, p. 15)

Compared to the transhuman we are naturally flawed because phusis proved to be incapable of offering us what we really deserve. Omnipotent Mother Nature "did the best she could" with the human being but that was not enough compared to what technology can offer us. Seen in Heideggerian terms, transhumanism is the human being as the violence-doer that reaches the apogee of his violence-doing: he releases all of his forces against phusis, no longer with the aim of making it reveal Being in beings, but in order to expose the flawed and uncompleted character of phusis, and therefore of Being Transhumanism, thus, arises as the absolute rebellion against phusis, as the demand for a new "natural" way of being.

4.4   Positionality as the Essence of Technology: Phusis and Human Positioning

Through transhumanism, technology, as we have seen so far, becomes a kind of (r)evolution, a new way of being, its goal consisting in founding anew humanity and the human being's essence on a ground very different from phusis' overwhelming sway. In modern society the advance of technology is the synonym of progress and civilization since it ceaselessly opens new paths for human beings' technical and personal development. Technology has made our life easier—at least some aspects of it—and more organized; geographical limits have vanished due to the internet. Technology, thanks to its breakthroughs over the centuries, has also served human life, taking into account all the advances in medicine, among other domains. Hence, it could be said that what transhumanism desires is the continuous enhancement of human life by technological means; an enhancement that would also seriously contemplate the moral issues technological progress may raise. According to Persson and Savulescu:

> However, it must not be overlooked that biomedical techniques of moral enhancement raise the same moral problem as all technological innovations: that of a proper application of them. In the case of techniques of moral enhancement, this takes the form of a bootstrapping problem: it is human beings, who need to be morally enhanced, who have to make a morally wise use of these techniques. Thus, the discovery of such techniques offers no foolproof way out of the danger zone into which humankind plunges deeper and deeper. (Persson and Savulescu 2010, p. 668)

Thus, technology's advance should always be followed by a deeper moral understanding of the dangers the former's development entails. Seen exclusively in this light, transhumanism could be well conceived as another phase in human history's technological development. Nonetheless, the difference, as we have also seen before, is that in transhumanism, for the first time, technology openly defies phusis, and the "will to life" becomes the only value guiding our every step. In transhumanism transcendence is no more, since all the values have been brought down to earth in order to serve the only possible reality which is nothing other than "will to power" as "will to life". Furthermore, transhumanism has finally managed to merge into one the organic and the mechanical, life and technology. Seen from this perspective, transhumanism emerges as the most adequate solution after Nietzsche's proclamation of the death of the God. By technologically enhancing human beings we can enhance life itself, and technology, instead of being considered as a tool, becomes an active part of our new bodily and mental existence. Nevertheless, all these technological breakthroughs, notwithstanding their undisputed potential significance, always come with a price. Glorifying life as the only and absolute value does not mean that we have finally understood the meaning of life; enhancing our bodily and mental capacities does not also imply that we will deepen our knowledge regarding our being and who we really are or how we are in this world.

Taking into account the goal of transhumanism—namely, to change our natural way of being and relating—the perspective through which we grasp and comprehend technology cannot be limited to the inventions and use of technological devices. The issue at stake is that transhumanism challenges the role of phusis itself. Conceiving technology as the Promethean blessing, while, at the same time, offering it the possibility to reshape our way of being proves that we have still a lot to understand regarding technology's essence and the potential dangers it may unleash.

Heidegger, in his interview with *Der Spiegel* magazine, made the following comments regarding technology[4]:

> Everything is functioning. That is precisely what is awesome {das Unheimliche}, that everything functions, that the functioning propels everything more and more toward further functioning, and that technicity{die Technik} increasingly dislodges man and uproots him from the earth. I don't know if you were shocked, but [certainly] I was shocked when a short time ago I saw the pictures of the earth taken from the moon. We do not need atomic bombs at all [to uproot us] – the uprooting of man is already here. All our relationships have become merely technical ones. (Heidegger 1981, p. 56)

Since everything is functioning well thanks to technology, why shouldn't we develop technology even further, as transhumanists advocate, making it part of our human nature, thus guaranteeing our prolonged bodily and even mental well-being? Because what is altered drastically through this almost perfect functioning is our relation with the world, with the beings that surround us. As Heidegger says, modern human beings are already uprooted: we don't need the destructive effects of an atomic bomb that would wipe out every possible ontic and ontological relationship, since technology has managed to become the core around which every potential relationship may be developed. Transhumanism brings technology to the next level due to the fact that it converts it to the only possible unconcealing act of both the human being and the world. Our point of reference, following transhumanism, is no longer phusis and its beings but exclusively the human being and the way the latter can be further developed and enhanced according to the standards set by the

---

[4] The translations of the words provided in brackets are my additions.

human being himself. The human being, through the means provided by technology, becomes the absolute center of every relationship developed in the world. Everything around us is becoming raw materials and resources ready-at-hand utilized in order to open the way for the arrival of the new enhanced transhuman. Our relation with the world now takes the form of an exclusive relation with ourselves, with the sole goal of further enhancing our own bodily existence and our lived experience. The world and the beings around us are now the proof that phusis has limited possibilities and that the human being is the only being that can rise to the challenge of becoming the new creator that will undo the wrong done to our species. The relation between human beings and phusis is now reaching a different level, where human beings are related only to the "ought to" of the becoming of our species; human beings' sole referential point is now the potential fulfillment of all those capacities that for millennia phusis has deprived us of. Transhumanism, through technology, is paving the way for "total mobilization"[5] for "the will to life".

Heidegger, in his lecture "Insight Into That Which Is", offers his own interpretation of technology,[6] directly relating it to Being and the way the latter is manifested in the modern world. Heidegger considers that we are witnesses of technology's unfolding of its own essence as "positionality (Gestell)"[7]—an event that distorts the way a human being relates to other humans, to beings and to the world in general.

One of the definitions of positionality presented by Heidegger is the following: "We now name the self-gathered collection of positioning [*des Stellens*], wherein everything orderable essences in the standing reserve, *positionality* [das Ge-Stell]" (Heidegger 2012, p. 31). Before further developing this definition we should first understand the significance of position for the German philosopher, since it is only by better grasping this that we can witness the (distorted) way in which positionality eventuates in our thinking, acting and being with people and beings. Hence, following Heidegger, we can better comprehend what position is if we see it in the light of the Greek word θέσις. Quoting Heidegger:

> What does the word θέσις say when we think it in a Greek manner? θέσις means positioning, placing, setting [*stellen*]. This positioning corresponds to Φύσις, so much so that it is defined by Φύσις, within the region of Φύσις, and from its relation to Φύσις. This points out that within Φύσις itself a certain θέσις-character is concealed. In the Greek world a chief distinction was expressed by the words φύσει and θέσει. It concerns what presences as such in the way that it presences. The distinction concerns the presencing of what presences, i.e., it concerns being. θέσει, θέσις is accordingly thought in relation to being. Thus the relationship between being and positioning was already announced in the first epoch of the history of being. (Heidegger 2012, pp. 59–60)

Thus, position, as θέσις, is directly related to phusis and being. Heidegger comments concerning this relation:

---

[5] *Total mobilization* is a term used by Ernst Jünger, in an essay bearing the same title, in order to describe the total conversion of life into energy.

[6] For a comprehensive account of Heidegger's philosophy of technology see Wendland et al. (2020).

[7] Positionality is the word Andrew Mitchell uses to translate "das Gestell", instead of the commonly used term "Enframing".

What is brought here forth in Φύσις is standing here [*herständig*] in unconcealment not through a human production, but rather through a bringing-here-forth of itself from itself. Bringing in the manner of Φύσις is now a positioning alongside, a positioning that sets up in unconcealment that which presences from itself. Φύσις, setting itself up in the unconcealed, is the letting presence of what presences in unconcealment. The letting presence of what presences is the being of beings. So construed, Φύσις, the emerging bringing-here-forth from itself, showed itself from early on to bear the character of a positioning that is not a human accomplishment. On the contrary, Φύσις first brings what presences as such to human production and representation by simultaneously giving unconcealment to humans and placing it at their disposal. (ibid., p. 61).

In this passage, we can see once again the recurring theme of the emerging activity of phusis as bringing beings forth into unconcealment. Beings, having been brought forth into unconcealment by phusis, radiate in their presence and are positioned by a positioning that sets them up and lets them truly be as phusis. Human beings, thanks to the gift of phusis and through the violence-doing of techne, can now grasp and represent beings, as well as produce things out of them, just like producing a statue out of stone. Phusis has both granted us beings' presence in their radiance and placed at our disposal the unconcealment of beings. Still, as Heidegger stresses, there is nothing in this first positioning that is our accomplishment: namely, an achievement of the human being. Being as phusis is granting us the first positioning as the gift of the unconcealment and only afterwards can the human being carry out his own positioning as production and/or misuse and exploitation. If there was not first the positioning of phusis, then the human being would be totally unable to position anything by himself. Quoting Heidegger:

Only when Φύσις reigns is θέσις possible and necessary. For only when there is something present that is brought about by a bringing-here-forth can human positioning, θέσις, then arrange upon such a presence (i.e., the stone) and out of this presence (stone) now something else that presences (a stone staircase and its steps), here among what is already present (the native rocks and soil). (ibid., p. 59)

What we can understand from the above passage is that the human being can only carry out the positioning because the emergence of beings has already taken place. As human beings we can never bring into unconcealment the stone, as phusis did, but through our human positioning we can let the stone present itself as a stone staircase, as a stone bridge, etc. Therefore, human positioning brings into a different kind of standing what has previously emerged from phusis. Our role as "positioners" is not to substitute phusis as emergence, but to bring beings into the light of a different standing by means of our own positioning. Thus, the marble can become a temple and the vineyard the wine that we dedicate to god. The human positioning of beings is of primordial importance because through this positioning much deeper changes take place, changes that can shift the destiny of our own being and Being itself, depending on our approach toward beings and their significance in our life and thought. Thus, we can either let beings truthfully be as unique things or transform them into mere raw material, degrading them to the status of replaceable cogs in a bigger machine, valued only for their utility and usefulness.

Therefore, the positioning carried out by the human being can either bring him closer to Being through his getting appropriated by it in the overwhelming sway of

Being as phusis, or pave the way for the eventuation of Being's distorted essence as positionality. The latter, namely this sort of human positioning, that is challenged forth by Being's distorted essence as positionality, is presented by Heidegger as follows:

> Men and women must place themselves in a work service. They are ordered. They are met by a positioning that places them, i.e., commandeers them. One places the other. He retains him. He positions him. He requires information and an accounting from him. He challenges him forth. Let us now enter into the meaning of this word "to position, place, set" so as to experience what comes to pass in that requisitioning through which an inventory arises [*der Bestand steht*] and is thus a standing reserve. To place, position, set means here: to challenge forth, to demand, to compel toward self-positioning. This positioning occurs as a conscription [*die Gestellung*]. The demand for conscription is directed at the human. But within the whole of what presences, the human is not the only presence approached by conscription. (Heidegger 2012, p. 26)

In this positioning, we witness how beings, human beings included, are ordered, commanded and retained. The terms conscription and requisitioning making their appearance here, bring forth a machine-like positioning wherein every being is ordered and strictly limited to its specific function. Beings, thus, are now challenged forth by Being in order to become mere tools for the sake of endless production. This whole process is compared to the creation of an inventory, where every being is transformed into an item, ready to be described, analyzed, categorized and stored, waiting for its future use. Beings become a standing reserve, losing the uniqueness and the radiance of their presence. The force that positions them in this state is stocking up on beings so as to guarantee their quantitative abundance, while, at the same time, it strips them of every qualitative characteristic they have, guaranteeing, in this way, their replaceability and the fading of their uniqueness as beings. The human being becomes in this way a mere tool of a power he cannot control, despite his illusion that he is still the master of his own will. The human being is actually answering a call when deciding to carry out positioning as requisitioning; nonetheless, he is not aware of the fact of his own answering and is led to believe that requisitioning is his own decision to act, his own way to question phusis and the world. Heidegger describes the human being's relation with requisitioning in the following way:

> The human himself stands now within such a conscription. The human has offered himself for the carrying out of this conscripting. He stands in line to take over such requisitioning and to complete it. The human is thereby an employee of requisitioning. Humans are thus, individually and in masses, assigned into this. The human is now the one ordered in, by, and for the requisitioning. Requisitioning is no human deed; in order for human effectiveness to cooperate each time in the requisitioning, as it does, it must already be orderable by this requisitioning for a corresponding doing and allowing. (ibid., p. 29)

But who is the one challenging forth the human being, requisitioning him and assigning to him the carrying out of positioning as requisitioning? According to Heidegger, the force guiding and ordering the requisitioning of beings and human beings is positionality. Quoting Heidegger:

> The word "positionality" names the essence of technology. Technology does not essence in the manner of a requisitioning and pursuing due to the technological process of building

and using an apparatus, something that still appears to us as a "framework" [*Gestelle*] in the sense of scaffolding and equipment. The essence of technology bears the name positionality because the positioning that is named in positionality is being itself [*das Sein selber*], but being at the beginning of its destiny had illuminated itself as Φύσις, as the self-emergent delivering that brings-here-forth. (ibid., p. 62)

Being itself is thus challenging forth the human being, requisitioning and transforming him into a tool whose only role is the further requisitioning of the rest of the beings. Being, that first gave a hint of its emergence in its concealment in the wondrous unconcealment of beings by means of the overwhelming sway of phusis, is now essencing as the requisitioning of beings, the same beings that could only emerge as phusis and through phusis.

But should we consider then that human beings have no share in deciding their own destiny? Are we just the puppets of Being's distorted essence, doomed to obey and follow a destiny that is not ours? The answer is no. As we have already seen, the human being is deinon because, along with the other beings, he belongs to the overwhelming sway of phusis. Being needs the site of the human being in order to be apprehended as phusis and the human being needs Being so as to meaningfully reveal the world and to bring Being into beings. Positionality, thus, is also a way of revealing:

The unconcealment of the unconcealed has already come to pass whenever it calls man forth into the modes of revealing allotted to him. When man, in his way, from within unconcealment reveals that which presences, he merely responds to the call of unconcealment even when he contradicts it. Thus when man, investigating, observing, ensnares nature as an area of his own conceiving, he has already been claimed by a way of revealing that challenges him to approach nature as an object of research, until even the object disappears into the objectlessness of standing-reserve. (Heidegger 1993, p. 324)

We human beings have the unique ability to reveal the world, to seek our being and meaning on earth. Through techne we decide to release our violence-doing in order to uncover Being in beings. Nevertheless, our act of revealing comes always as an answer to the claim made on us by Being itself. Through phusis and as phusis we were granted the gift of the wondrous appearance; through positionality we are now challenged forth as the requisitioners, the masters of the earth and phusis. But every single act of revealing comes as an answer to the claim made by Being. Being challenges us ceaselessly, releasing its violent overwhelming sway against us. Still, when positionality takes place, Being slides toward the oblivion of its own truth: it directs its violent sway against its own essence, and is thus distorted and alienated from its inception as phusis.

Having presented the essence of technology as positionality we can now better understand the challenges set forth by transhumanism. Transhumanism represents the twofold challenge set by Being as positionality: that is, the replacement of the inceptual positioning of phusis by means of technological enhancement—thus assigning to the human being reign over the overwhelming sway of emergence—and the requisitioning of the human being himself, through the latter's instrumentalization, to fulfill the greater destiny of the "will to life". Technology's primary focus is no longer the variety of beings surrounding us and their exploitation for the benefit of humanity;

technology as positionality is now converting the human being into the new raw material, the new source of energy that will enable the arrival of the transhuman. As a consequence of positionality, and in the name of a "better life", the human being lets technology enter into his own being, to become one with him, hoping for a brighter and more technologically enhanced future that would guarantee a stronger and healthier physical existence. But the price is high: accepting our being requisitioned by positionality at the altar of "life" we are turning against our own selves since we are deciding to accept the alteration of our own bodily and mental capacities for the sake of an upgraded version of ourselves. In other words, the omnipotent subject of the modern society, having downplayed the role and the importance of phusis, is now degrading its own existence, by projecting a potentially much better version of itself. Positionality, as reflected in transhumanism, makes us consider our own existence insufficient when compared to what we could be, thus converting technology into a sine qua non element of humanity's future. Transhumanism, having characterized our natural way of being as flawed, now turns to technology so as to take the scepter of creation from phusis. This, however, leads to a very contradictory conclusion: namely, that the enhancement of life can only take place through the absolute dominance of artificiality.

4.5 Crossing the Last Frontier: Death in Transhumanism and Heidegger's Thought

The last part of this chapter is dedicated to the delicate issue of death in transhumanism and Heidegger's thought. It is no secret that overcoming death has become a sort of "Holy Grail" for transhumanists: "The true goal of transhumanism is the defeat of aging and death. It is a Promethean ambition but increasingly we see steps in that direction (Broderick 2013, p. 434)". So says Damien Broderick, choosing the mythical figure of the titan Prometheus to highlight the crucial role of transhumanism. Nick Bostrom, in his short story "The Fable of the Dragon-Tyrant" depicts death as the bad dragon and optimistically foresees the dragon's death, the death of death, at the hand of a stronger transhuman. As we have seen before, the fact that death is something natural—a tragic event, of course, but, still, natural—is of little importance for transhumanism. Quoting Bostrom:

> Transhumanists insist that whether something is natural or not is irrelevant to whether it is good or desirable (…) The quest for immortality is one of the most ancient and deep-rooted of human aspirations (…) If death is part of the natural order, so too is the human desire to overcome death. (Bostrom 2005)

Death, we may deduce, is deprived of its ontological content in transhumanism, while the "will to life" seems to be overburdened by the latter. Let's see, for example, how the process of mind uploading, a part of Transcendent Engineering, and another way to defeat aging and death, is described in the following passage:

Therefore I have always been interested in mind uploading and I consider it as the "Holy Grail" of tranhsumanism: let our minds break free of our biological brains and bodies, and we will be free to roam the universe and grow beyond limits as "software angels." (Prisco 2013, p. 235)

The "Holy Grail", "software angels" and the release from every possible bond, even biological ones, make their appearance, letting a "mysticism" of life emerge, even when life is no longer conceived as we currently understand it. Life becomes a quest, a crusade, the limitless universe wherein we roam, while death is the mere enemy, the obstacle, and the no longer absolute limitation of our existence, bodily or not. Life has to go on indefinitely since it is life that lies at the core of transhumanism's philosophy and meaning. While the "will to life" is glorified and even mystified, death is restricted to its role as a malfunctioning. Quoting Brian Wowk, "This reminds us that death is not when life turns off. Death is when the chemistry of life becomes irreversibly damaged (Wowk 2013, p. 221)", and "death only happens when biochemistry becomes irreversibly damaged, and "irreversibility" is technology-dependent. (ibidem.)" Death's irreversibility, thus, being technology-dependent, could one day be overcome through the means of cryonics and the uploading into software of the contents of the minds, among other methods that some decades ago would have been considered far-fetched and pure creations of the imagination. If transhumanism manages to progress and start implementing its technological agenda, if death finally becomes an option and not an inevitability, wouldn't it be a good option for human beings to choose their own moment and circumstances of death? We could go on living and when, finally, for our own reasons, we decide that we no longer want to live, either in bodies or as "software angels", then we could finally let go and rest.

The ontological issue that emerges here is that perceived like that, death is no longer an inevitability, the unavoidable and uncontrollable end of our own finite existence, but a simple option, among many others. Death loses its significance as "the possibility of the impossibility of any existence at all (Heidegger 2001, p. 307)" and becomes just another possibility. But the transformation of death from impossibility into a choice depending on us, brings forth the issue of care and decision. Namely, as long as we live, knowing that our life will inevitably end, our actions and our decisions are limited. Nevertheless, this limit, instead of making us feel vulnerable and weak, gives our actions and thought the significance they deserve. Our decisions shape us, form who we are and when the moment of our death finally arrives our whole life would be a pathway we ourselves have decided to open and follow. The same happens regarding our care about beings and human beings around us; we decide to dedicate to them our time, knowing that our days on earth are not infinite. Time is precious, because it is limited and makes our caring about others unique. We decide to care about the others, we decide to care about our own being embarking upon long "ontological journeys" because we understand the importance of sacrifice. We sacrifice our time because we want to reach a deeper and more authentic mode of existence.

Transhumanism, on the other hand, offers us the gift of limitless time and decisions. Simply said, if transhumanism finally manages to overcome death, then everything else happening in our lives could be reversible too. There will be no longer any impossibility of our possibilities since the limit granting us time, death, will no longer halt our breakthrough to eternity. Howbeit, if everything is reversible then our decisions as well will essentially lose their significance and gravity. How can we make an important decision knowing that we will have infinite time to undo every possible drawback provoked by it? What is the meaning of the decision if it can hardly leave a mark on our lives? How, finally, can our decisions shape our personality and mold who we are if they can be repeated once and again? Death, thus, may be a limit and a boundary; it is inside this boundary, however, that time, care and essential decisioning take place and flourish as such.

Still, even in a case where we finally conquer and control death, we would never become immortals. Because understanding what immortal means, requires a much more profound comprehension of the essence of mortality: a comprehension that is founded on our experiencing death ontologically, meaningfully and not purely ontically, as the obstacle that does not permit the "eternal recurrence of now" of the lived experience. In a few words, death, as the shattering of possibilities and not its transhumanist distortion, gives meaning, space, time and destiny to our lives: it unfolds Being inside our world. Quoting James Demske:

> The concept of death as a shattering against being points up the essential connection between death and man's position as the lighting-up-place of being. In order that being may come to presence, appear, or be lighted up, man must act as the staging area where being transforms itself into beings, or as the breach through which being breaks into history. (Demske 1970, p. 144)

Our getting shattered by death and by the overwhelming sway of Being is the open space for Being's becoming history. We, human beings, are the breach through which Being makes manifest its presence in the world. Surrendering ourselves to Being and accepting death, we accept the uncanny yet wondrous breakthrough of Being inside our lives; a breakthrough, that as we will now see, is never limited to pure presence and/or objectification of beings. Through death Being makes itself manifest via its non-presence; a non-presence, however, that never becomes nothingness.

This unfolding of Being is not always a mystical experience for the few and for the initiated; our relation with the other human beings and with beings in general, as well as the way we approach our world and our reality, make Being manifest in our thought and action. Heidegger presents this manifestation of Being through an act of human positioning: namely, the creation of the "death-tree (Totenbaum)":

> The carpenter in the village does not complete a box for a corpse. The coffin is from the outset placed in a privileged spot of the farmhouse where the dead peasant still lingers. There, a coffin is still called a "death-tree" [*Totenbaum*]. The death of the deceased flourishes in it. This flourishing determines the house and farmstead, the ones who dwell there, their kin, and the neighborhood. (Heidegger 2012, p. 25)

The above passage gives us a clear idea of how death should be understood and conceived according to Heidegger. Heidegger does not speak about "boxes for corpses" but of "death-trees", thus offering a radically different approach concerning our death and its significance in a community and tradition. Death "flourishes" inside the "death-tree", becoming a point of reference for all those who surround it, thus essentializing our relation with it. When the coffin is seen as the "death-tree" wherein death flourishes, then a simple material production—a coffin, in this case—becomes the symbol of our own mortality and the proof of the mortality of the deceased. The deceased has fulfilled his essence as mortal through his dying, while the family and the community are the mortals who are still journeying toward death.

We are mortals because we are always related to death and being mortals we are always directly related to Being. Our mortality is the scar Being lets on our being during the overwhelming sway of its emergence as phusis. When we grasp and accept mortality as our only way of being, Being makes its appearance manifest to us. Following Heidegger,

> The mortals are the humans. They are called the mortals because they are able to die. Dying means: to be capable of death as death. Only the human dies. The animal comes to an end. It has death as death neither before it nor after it. Death is the shrine of the nothing, namely of that which in all respects is never some mere being, but nonetheless essences, namely as being itself. Death, as the shrine of nothing, harbors in itself what essences of being. As the shrine of the nothing, death is the refuge of being. The mortals we now name the mortals—not because their earthly life ends, but rather because they are capable of death as death. The mortals are who they are as mortals by essencing in the refuge of being. They are the essencing relationship to being as being. (Heidegger 2012, p. 17)

Only we, mortals, when we understand and concernfully approach our own mortality, are capable of death as death. This is why the "death-tree" is not a "box for corpses". But how can we understand the cryptic statement that death is the "shrine of nothing" and the "refuge of being"? This can only be understood when placed in the context of a relation between the human being and his death, between the manifestation of Being and its concealment in death. Quoting Andrew Mitchell:

> The shrine thus performs a preservative function. It keeps the dead one from dying, or, more appropriately, it keeps the dying one from their death. There is neither pure negation nor annihilation here thanks to the shrine. What is enshrined is not nothing, though this is indeed the shrine of the nothing. It is maintained in a state "between" being and nothing. (Mitchell 2015, p. 234)

In this passage, Mitchell presents death from an ontological perspective that relates the mortals with Being and Nothingness. Of course, people around us will die and sadness or even frustration may arise from their loss. We may feel helpless when facing death and having no control over it fills our life with anxiety and dread. The dark shadow of Nothingness is cast all around us when thinking about death and the hope for a peaceful beyond can quickly turn into desperation and trembling. But all of these feelings cannot be converted into a desire to remove death from our life because, as Mitchell thoughtfully explains, through death, the dead one, the one who has become mortal through his death, is "between" Being and nothing. The fact that the deceased is not present does not mean that he has vanished into nothingness. Were it so, then no possible relation could ever be developed with him due to the fact of his total and absolute removal. When we go to a cemetery, however, we can mourn, "speak" with the dead, and open our hearts and thoughts in front of their grave. They are there for us, but not as an objectified presence. They are and they are not: they are the between that lets them keep being even if they are not by our side anymore. Seen in this perspective, we can also now understand the reason why Heidegger makes reference to death as the "refuge of being". Quoting Mitchell again:

> Death is the refuge of being because it relieves it of the obligation to unqualifiedly be, to be present. Death grants being concealment. It brings a protection to being just as much as to the nothing. Death gives weight to the nothing and alleviates being. (ibid., p. 235)

Being finds refuge in death because death unburdens it from the obligation of the constant presence. Being still *is*, even if it is not present: in the same way, the dead are far away from us, but still *are* in our lives. Transhumanism demands constant presence: one way or another we must always be present, either through our bodily existence or as software programs and uploaded consciousness. Death is perceived as the absolute enemy because it conceals Being: it degrades the idea of beings as ready-at-hand present beings, offering concealment and refuge. This being there while not being present is unacceptable for transhumanism and this is why death's presence will never be accepted or justified in this movement. Death is not, thus, the enemy, to banish from our lives. We are mortals because we will die and death will hurt us because it is the threshold between Being and nothingness. It pervades our lives, it brings us fear, anxiety and pain, but, still, death makes us what—and who—we are as mortals. Thus, mortality, in Heidegger's thought, is not to be understood as the unbreakable chain of a destiny that dooms us to the fate of a limited existence; quite the contrary, it becomes a bridge in the depths of the abyss of Being.

Finally, death stands forth as the proof of a differentiation that we human beings will never be able to overcome: namely, the differentiation between the overwhelming sway of Being and beings, between the world and things. Quoting Heidegger:

> For world and things do not subsist alongside one another. They penetrate each other. Thus the two traverse a middle. In it, they are at one. Thus at one they are intimate. The middle of the two is intimacy—in Latin, *inter*. The corresponding German word is *unter*, the English *inter-*. The intimacy of world and thing is not a fusion. Intimacy obtains only where the intimate—world and thing—divides itself cleanly and remains separated. In the midst of the two, in the between of world and thing, in their inter, division prevails: a *dif-ference*. (Heidegger 1971, p. 202)

World and things, Being and beings are separated and the "between" is "the pain that has turned the threshold to stone (ibid., p. 204)". As we have seen in the first part, the human being as deinon is one among the other beings that have been brought forth by the overwhelming sway of phusis; but being in this sway is being in pain, since Being, phusis, cannot be conquered, no matter how violent deinon might become. Our violence-doing will always be shattered by the violent sway of phusis and it is precisely this shattering, the defeat of our violence-doing, that allows us to be the site of the revelation of Being. The pain of the between of this differentiation between Being and beings is compared by Mitchell to a wound:

> But for the intimacy of the between there must be those capable of such intimacy, disposed to such mediation. Such things are relationally pouring through the world, split onto the world, cut open. This cut runs through the thing exposing it, cutting it open to the sky. Any notion of integrity must be abandoned. What can be intimate are the cut open, the wounded. Being is the exposure of this wound, the beyond called together by this cut. (Mitchell 2015, p. 303)

Life, thus, is an open wound because as long as we live we will never be able to heal the cut that separates us from Being. Beings emerge around us and we emerge among them and we are all exposed to the overwhelming sway of Being. Our exposure is a wound that will never close, because its healing, while we are still alive, would demand control over Being, over the one who keeps this wound alive through the differentiation. As long as we live we are in pain and we are dying, recalling Heidegger, but this pain and suffering are the only way through which Being appears to us, through its non-appearance. It is in our hands to let the projection and apprehension of death temporalize and spatialize our own life. Giving death the significance and importance it deserves, life, even if ontically limited, becomes a true life. Denying life the event of death would distort life's essence, converting it into a blind "will to life", aiming at nothing else than its own endless expansion. Life, thus, drawn violently out of its own limits, would become like Phaethon, who, unable to control Helios' chariot, scorched everything in his path before being killed by Zeus' thunderbolt.

4.6 Concluding Remarks: As Mortals Gaze at Death

Death, as we have seen above, grants the meaning of life by showing us the inaccessibility of Being's domain. Still, Being only manifests itself to us through its inaccessibility, through mortals' not being able to control the origin of the overwhelming sway. Death is a limit and a frontier: not only because it limits the span of our life's duration, but also because it unfolds the site of our own being, which is nothing other than our mortality. If we were to shatter this limit, we would never become immortal, because the immortal is only the one who stands against the mortal, through the relation of confrontational intimacy. If transhumanism manages to overcome the "obstacle" of death this would not erase suffering and pain from our lives: it would only deprive them of their significance and meaning. We would be lost inside the vortex of the "will to life", endlessly going forward toward the unknown, not knowing who else remains to be beaten after death. Taking away death, we do not offer ourselves the gift of life: we just change life into the "eternal recurrence of now", where everything is open and revealed to us through positionality. The attack on death is an attack on the meaning and the essence of life, depriving the latter of its relational existence. Tearing apart the relation between death and life, we get caught in a between that is no longer between since there is nothing to fill the gap. The "will to life" will keep on trying to become the new overwhelming sway, the new phusis, the transhuman's Being, but every effort will be in vain because, even without death, Being will still remain the overwhelming sway, the only possible phusis. transhumanism, no matter how far it advances, will never become Being because, even if humanity is the "half-baked beginning" of Being's sway, still this sway is the origin and the origin is irreplaceable and unrepeatable. What comes after the beginning will always be an imitation, a distortion of the beginning—notwithstanding its greatness—it will never be another origin.

References

Ansell Pearson, Keith. 1997. *Viroid Life: Perspectives on Nietzsche and the Transhuman Condition.* London: Routledge.

Blok, Vincent. 2014. Reconnecting with Nature in the Age of Technology: The Heidegger and Radical Environmentalism Debate Revisited. *Environmental Philosophy* 11 (2): 307–332.

Blok, Vincent. 2017. *Ernst Jünger's Philosophy of Technology.* New York: Routledge.

Bostrom, Nick. 2001. Transhumanist Values. https://www.nickbostrom.com/ethics/values.html. Accessed 24 July 2020.

Bostrom, Nick. 2003. The Transhumanist FAQ. World Transhumanist Association.

Bostrom, Nick. 2005. The Fable of the Dragon-Tyrant. https://www.nickbostrom.com/fable/dragon.html. Accessed 15 July 2020.

Broderick, Damien. 2013. Trans and Post. In *The Transhumanist Reader: Classical and Contemporary Essays on the Science, Technology, and Philosophy of the Human Future*, ed. Max More and Natasha Vita-More, 430–437. New Jersey: Wiley.

Demske, James. 1970. *Being, Man, and Death: A Key to Heidegger.* Lexington: The University Press of Kentucky.

Heidegger, Martin. 1971. *Poetry, Language, Thought.* Trans. Albert Hofstadter. New York: Harper and Row Publishers.

Heidegger, Martin. 1981. *Heidegger, The Man and the Thinker*, ed. T. Sheehan. London: Routledge.

Heidegger, Martin. 1991. *Nietzsche: Volumes III and IV.* Trans. Joan Stambaugh, David Farrell Krell, and Frank A. Capuzzi. New York: HarperCollins Editions.

Heidegger, Martin. 1993. *Basic Writings*, ed. David Farrell Krell. San Francisco: Harper.

Heidegger, Martin. 2001. *Being and Time.* Trans. John Macquarrie and Edward Robinson. New Jersey: Blackwell Publishers.

Heidegger, Martin. 2012. *Bremen and Freiburg Lectures.* Trans. Andrew J. Mitchell. Indiana: Indiana University Press.

Heidegger, Martin. 2014. *Introduction to Metaphysics.* Trans. Gregory Fried and Richard Polt. Yale: Yale University Press.

Mitchell, Andrew, J. 2015. *The Fourfold: Reading the Late Heidegger.* Illinois: Northwestern University Press.

More, Max. 2013a. A Letter to Mother Nature. In *The Transhumanist Reader: Classical and Contemporary Essays on the Science, Technology, and Philosophy of the Human Future*, ed. Max More and Natasha Vita-More, 449–450. New Jersey: Wiley.

More, Max. 2013b. The Philosophy of Transhumanism. In *The Transhumanist Reader: Classical and Contemporary Essays on the Science, Technology, and Philosophy of the Human Future*, ed. Max More and Natasha Vita-More, 3–17. New Jersey: Wiley.

Plato. *Protagoras.*

Persson, Ingmar, and Julian Savulescu. 2010. Moral Transhumanism. *Journal of Medicine and Philosophy* 35: 656–669.

Philbeck, Thomas. 2014. Ontology. In *Post- and Transhumanism: An Introduction*, ed. Robert Ranisch and Stefan Lorenz Sorgner, 173–183. Frankfurt am Main: Peter Lang.

Polt, Richard, and Gregory Fried, eds. 2001. *A Companion to Heidegger's "Introduction to Metaphysics."* Yale: Yale University Press.

Prisco, Giulio. 2013. Transcendent Engineering. In *The Transhumanist Reader: Classical and Contemporary Essays on the Science, Technology, and Philosophy of the Human Future*, ed. Max More and Natasha Vita-More, 234–240. New Jersey: Wiley.

Ranisch, Robert, and S. Lorenz Sorgner. 2014. Introducing Post- and Transhumanism. In *Post- and Transhumanism: An Introduction*, ed. Robert Ranisch and Stefan Lorenz Sorgner, 7–27. Frankfurt am Main: Peter Lang.

Savulescu, Julian. 2005. New Breeds of Humans: The Moral Obligation to Enhance. *Reproductive BioMedicine Online* 10: 36–39.

Wendland, Aaron James, Christopher Merwin, and Hadjioannou Christos, eds. 2020. *Heidegger on Technology*. London: Routledge.

Wowk, Brian. 2013. Medical Time Travel. In *The Transhumanist Reader: Classical and Contemporary Essays on the Science, Technology, and Philosophy of the Human Future*, ed. Max More and Natasha Vita-More, 220–226. New Jersey: Wiley.

Georgios Karakasis received his PhD in Philosophy from the University of the Basque Country (UPV-EHU). He got his degree in Political Sciences and Public Administration at the National and Kapodistrian University of Athens and holds two Master's degrees in Humanities (UPV-EHU) and Phenomenology of Terrorism (University of Granada). He has taught Constitutional Law and Gender Equality at the University of the Basque Country and his main research interests include the later thought of Martin Heidegger, the philosophy and ideology of fascism as well as the thought of Julius Evola.

Chapter 5
Mens Humana in Corpore Humano—Body-Hacking the Human Experience

Robin Markus Auer

Abstract While there is a remarkable range of views on what defines the human experience, what seems relatively uncontroversial is the conviction that it is fundamentally dependent on the ways in which we interact with the world as embodied agents. If we take this position seriously, we must acknowledge that any changes made to the physical setup of individual humans (ontogenesis), or to our setup as a species (phylogenesis), must necessarily also result in a shift in individual, shared experience of human life. Studies have already presented evidence towards changes in behaviour, cognitive capabilities due to technologically enhanced ways of interacting with the world, an effect also observed in connection with cultural tools such as notational systems. Novel human-world interfaces are therefore likely to introduce new ways of experiencing the world. Any such fundamental changes can have critical real-world impacts, resulting in a loss of common, shared experience. It is, therefore, crucial to consider how such changes affect the social fabric of societies: the more diverse humanity becomes, the more important issues of equality, participation will be. Therefore, the challenges of such a development are threatening the functioning, cohesion of human society, rather than humanity as such.

5.1 Embodying Transhumanism

Transhumanism is often perceived as a fundamental threat to humanity, not only in terms of how certain technologies might work towards an eventual extinction of the species homo sapiens as we know it but also in terms of challenging the very foundations of what it means to be human. And while prominent transhumanists try to refute this claim of 'the world's most dangerous idea' (Fukuyama 2004, 42–43), there is a fundamental unease related to the very term 'transhumanism', and to the

R. M. Auer (✉)
Institut für Anglistik und Amerikanistik, 38106 Braunschweig, Germany
e-mail: robin.auer@tu-braunschweig.de

closely related posthumanism.[1] Transhumanism is a fascinating movement in that most of the attention it gets is derived from dystopian scenarios,[2] while most of its proponents insist that it is a fundamentally eutopian endeavour (Hauskeller 2012). Interestingly though, the problems that they claim to want to fix are not the ones that occupy most people's minds, exposing it to accusations of elitism. And while it is true that detrimental effects are likely to be socially mediated, we have to be aware that the same is true for the opportunities that come with technology. It is not only possible but indeed overwhelmingly likely, that an implementation of the transhumanist agenda would result in increased inequality and civil unrest.

The aim of this chapter is to present a pluralist and inclusivist perspective on some of the core issues related to transhumanism, as well as to argue that what has come to be known as body-hacking, while certainly altering an individual's subjective experience, does not necessarily affect to what degree an individual should be categorised as human, especially as the notion of the human is notoriously flexible and constantly undergoing change, reflecting the constant changes within human societies that are mirrored in the development of the mosntrous (Braidotti 1996). A further point that will be made is that while embodiment theory is a central pillar for many of the claims of transhumanists, it is often neglected when it comes to potential obstacles or inconsistencies. As a result, this chapter will argue that the term posthuman should be abandoned,[3] that transhumanism does not pose a threat to humanity (as either species or abstract concept), but rather to the very functioning and cohesion of human society, and that as a result, the transhumanist movement needs to acknowledge its political dimension more readily. Furthermore, it will be argued that in order to present an actual progressive movement, transhumanism requires a re-conceptualisation within a decidedly democratic framework, or—in other words— needs to shed its libertarian tendencies. In a way, this chapter is concerned with where transhumanism is currently situated as well with how it needs to proceed from now onwards.

We will, therefore, begin by taking a closer look at the semantic field surrounding trans- and posthumanism and what it reveals about transhumanism as a movement, in order to then move on to a closer examination of transhumanism's links to theories of situated cognition (focussing mostly on embodiment), followed by a discussion of why, from a philosophical perspective, transhumanism does not pose a fundamental risk to humanity in both of the senses specified above (as a species and as a concept). In a final step, the chapter takes a closer look at the challenges transhumanism

---

[1] In fact, this chapter argues that posthumanism and transhumanism are more closely related than both posthumanists as well as transhumanists usually care to concede. While I agree with claims made by transhumanists such as Sorgner, who argues that '[metahumanism, posthumanism and transhumanism] have to be clearly distinguished' (2016, 66), it seems necessary to point out that they nonetheless have to be conceptualised and comprehended within a common framework, and more importantly, operate within society in close conjunction.

[2] For an overview of the connection between transhumanism and modern renderings of apocalyptic scenarios, see Hughes (2012).

[3] This solely applies to the adjective posthuman, not to posthumanism.

presents to human society and societies, and why it is crucial to draw attention to transhumanism's political dimension.

5.2 Transhumanisms

While there is a crucial historical dimension to transhumanism and its relation to posthumanism, this section of the chapter will approach them from a linguistically informed philosophical perspective. For a movement that has been pushed so forcefully into public debate (where intuition and deduction overrule informed positions and nuance is consequently almost always lost), transhumanism remains notoriously difficult to pin down. Indeed, it would be far more accurate to speak of *transhumanisms*. This is due partly to a plurality of views and perspectives within the movement itself, which makes it a conglomerate of small movements rather than one coherent movement and rather complicates matters, as transhumanism already is part of a field of closely related and intertwined concepts. As Ferrando (2013) writes.

The label "posthuman" is often evoked in a generic and all-inclusive way […] creating methodological and theoretical confusion between experts and nonexperts alike. "Posthuman" has become an umbrella term to include (philosophical, cultural, and critical) posthumanism, transhumanism (in its variants as extropianism, liberal and democratic transhumanism, among other currents), new materialisms (a specific feminist development within the posthumanist frame), and the heterogeneous landscapes of antihumanism, posthumanities, and metahumanities. The most confused areas of signification are the ones shared by posthumanism and transhumanism. (26).

Even transhumanists acknowledge that 'what exactly transhumanism represents and which cultural, ethical, political, artistic and philosophical views it is associated with is not yet common knowledge' (Sorgner 2016, 65). Ferrando also points out that 'within the transhumanist debate, the concept of posthumanism itself is interpreted in a specific transhumanist way, which causes further confusion in the general understanding of the posthuman' (Ferrando 2013, 27), complicating matters further. It would seem that there are two competing notions of the posthuman, one claimed by its namesake posthumanism, and the other by transhumanism, which, due to the transitory nature it ascribes to humans in the transhuman age, can only envisage the result of the transition (the posthuman) rather than a transhuman. The transhuman is 'transitional to the posthuman' (Hughes 2012, 761), with the posthuman framed in a specifically transhumanist way. This is not necessarily a problem, but it increases the likelihood of misunderstandings and can also cause conflict between the two camps. Another reason for said confusion is the terminological ambiguity that is blessing and curse at the same time (in that it conjures up all kinds of fantastical promises that transhumanism then struggles to fulfil). As the arguments made in this chapter are of a theoretical and general nature, and a complete classification of subtypes of trans- as well as posthumanism, would more than exhaust its limits, we will now proceed with a short analysis of the semantic field and its implications, and will therefore necessarily engage with individual ideas rather than coherent conceptualisations of transhumanisms.

Dissecting the meaning of the terms trans-human-ism as well as post-human-ism according to their morphology primarily depends on the order of the affixes: there is a major difference between trans-humanism and transhuman-ism in that the former relates to humanism whereas the latter transforms the notion of a 'transhuman' or a 'transhuman' stage into a movement or ideology. While there is certainly considerable overlap between the two forms of transhumanism, they each possess distinctive features, and proponents of transhumanism usually tend towards either rather than both versions. It is also unsurprising that the latter version is more often associated with an almost religious zeal, while the former has a markedly philosophical element and is indeed more closely related to posthumanism, which positions itself as a philosophical perspective.

Posthumanism is morphologically similar to transhumanism in that there are also two possible derivations for the word posthumanism. There is the relatively straightforward post-humanism, a critique of humanism, and there is posthuman-ism, which is centred on the notion of the post-human. As we have already noted, the term posthuman is claimed by both posthumanists as well as transhumanists. In its transhumanist reading, the indebtedness to the Nietzschean 'Übermensch'[4] (super-human) is rather obvious, even though the term suggests a more objective perspective linguistically (with questionable success). According to Sorgner, '[t]he posthuman would thus be a being that no longer belongs to the human species, but instead presents an evolutionary successor [or improvement][5] to humanity' (2016, 18). In terms of an evolutionary perspective, the assumption underlying the transhumanist conception of the posthuman seems to be based on a misinterpretation of the theory of evolution[6]: for one, it seems to imply a kind of teleology of evolution in which something better ('improved') supplants that which has become obsolete through its arrival. As evolutionary biologists are keen to point out, this is not at all how evolution usually proceeds. Furthermore, species generally tend to coexist and develop alongside each other, with new ones developing and some going extinct. Any talk of successors is a matter of interpretation and not backed by evolutionary biologists. Also, it apparently needs to be pointed out that evolution is a biological process that is concerned with phylogenesis rather than ontogenesis. Evolutionary changes do not happen to an organism but to a species. So when Ferrando writes that 'the movement of transhumanism problematizes the current understanding of the human not necessarily through its past and present legacies, but through the possibilities inscribed within its possible biological and technological evolutions' (2013, 27), she falls prey to the

---

[4] This is contested by Bostrom, who only finds 'surface-level similarities with the Nietzschean vision' (2005, 4), but Sorgner concedes that this tendency of transhumanists to distance themselves from Nietzsche is due to his stigmatisation rather than actual conceptual differences (2016, 111).

[5] The German 'Weiterentwicklung' carries positive connotations and within a technophile context clearly implies 'improvement'.

[6] Or rather, it seems to commit the mistake of adopting something as a metaphor and then consequently taking the metaphor literally, which is quite a common fallacy at the meeting point between sciences and humanities.

same fallacy. There is no such thing as technological evolution. 'Revolution' more aptly describes the process by which technology advances. She is right, though, to point to the main difference between the two conceptions of posthuman employed by transhumanism and posthumanism in that transhumanism problematises the human through perceived possibilities while posthumanism does so on the basis of its history and current form.

5.3 Human by Experience, or Why Transhumanism is not a Threat

As we have seen, both transhumanism, as well as posthumanism, or indeed all of what Sorgner refers to as the 'beyond-humanism-discourse' (2016, 67), conceptually engages with humanism. At the same time, however, the more fundamental question of what defines humanity as such lies at the very core of any treatment of transhumanism as well as posthumanism, as it affects our understanding of the posthuman. In line with the main objectives of this chapter, we will leave aside humanism and concentrate on the human instead. More precisely, we will take a look at the transhumanist and posthumanist deconstruction and criticism of the human (both in terms of explicit positions and implicit assumptions), as well as offering a tentative suggestion for a more pragmatic notion of the human in the context of the wider 'beyond-humanism-discourse'.[7]

Following Ferrando, posthumanism is also 'a post-anthropocentrism: it is "post" to the concept of the human and to the historical occurrence of humanism, both based, as we have previously seen, on hierarchical social constructs and humancentric assumptions.' (2013, 29) In contrast to transhumanism, 'the posthuman overcoming of human primacy, though, is not to be replaced with other types of primacies (such as the one of the machines)' (2013, 29), which is due to posthumanism's rejection of 'frontal dualism or antithesis, demystifying any ontological polarization through the postmodern practice of deconstruction' (Ferrando 2013, 29). This is all commendable, but I think it misses the crucial point, which is that the human perspective and label derive most of their importance and centrality as a matter of epistemology rather than ontology. The human perspective is central to our thinking because it is our perspective, not because it is in any way ontologically superior to other perspectives. As such, the anthropocentric perspective is already situated and contextualised pre-posthumanism, implicitly so at least since Darwin's *Origin of Species*.

So when Ferrando contends that 'posthumanism stresses the urgency for humans to become aware of pertaining to an ecosystem which, when damaged, negatively affects the human condition as well […] located within an extensive system of relations' (2013, 32), this is arguably not opposed to anthropocentrism but rather

---

[7] This is partly driven by the conviction that it is usually better to reclaim a term than to abandon or simply reject it.

to anthropocentrism gone bad. Anthropocentrism is in no way whatsoever inherently committed to ignoring the situatedness of humanity. In fact, the argument for acknowledging the need to account for the entire ecosystem and its interdependencies is more urgent (to us as humans) within an anthropocentrism framework. Posthumanism, in turn, runs the risk of abandoning a limited but situated and contextualised perspective for an ill-fated attempt at an abstract perspective or even non-perspectivism. If what I have here described is also what Ferrando means by saying that:

> Posthumanism is a post-centralizing, in the sense that it recognizes not one but many specific centers of interest; it dismisses the centrality of the centre in its singular form, both in its hegemonic as in its resistant modes. Posthumanism might recognize centers of interests; its centers, though, are mutable, nomadic, ephemeral. Its perspectives have to be pluralistic, multilayered, and as comprehensive and inclusive as possible. (2013, 30)

then I fail to see what posthumanism claims to add to postmodernism. Ironically, it would seem that (subverting its own agenda) posthumanism is a more anthropocentric postmodernism. Posthumanism, in this view, is not so much an autonomous philosophy as it is a specific perspective within postmodernism whose advantage is 'allowing one to envision post-human futures which will radically stretch the boundaries of human imagination' (Ferrando 2013, 30) in a more focussed way.

Let us now take a closer look at transhumanism, which has a markedly more negative perspective on the human, which tends to undercut the genuine contributions it could make. This negative perspective is sometimes explicitly expressed, but more often implied in the choice of words and the underlying attitudes. According to one of its most prominent proponents, '[t]ranshumanists view human nature as a work-in-progress' (Bostrom 2003, 493), which is hardly a promising start. This is not to say that the claim is wrong (humanity is arguably changing constantly and at some speed). It is in the way the position is framed that the negative view is revealed. A 'work-in-progress' is not something that is generally subjected to change and thus not fixed, but something that is still faulty, unfinished, not yet perfected. Bostrom himself manages to express a similar sentiment somewhat more objectively when saying that '[a] common understanding is that it would be naïve to think that the human condition and human nature will remain pretty much the same for very much longer' (Bostrom 2001), but carries strongly Nietzschean undertones nonetheless.

Sorgner presents the difference between Darwin's theory of evolution and Nietzsche's focus on striving for power as a minor difference, taking survival as but a narrower aspect of power. While this may even be true as far as Nietzsche's criticism of Darwin's theories is concerned, we must acknowledge that this presents a fundamental shift towards a surprisingly humanist and teleological perspective that is not present in Darwin.[8] This is one of the aspects that makes transhumanism crucially

---

[8] Nietzsche's criticism of the focus on survival is derived from his anthropology. In other words, while Darwin's theory of evolution allows us to explain and understand human behaviour as a specific form of more general principles, the opposite is true for Nietzsche's understanding of evolution, which requires the superimposition of an abstract ideal of power onto evolutionary processes, implying a qualitative shift towards a normative understanding of evolution and adaptation.

Nietzschean and non-Darwinian in that it presumes a qualitative dimension of evolution. This is not only surprisingly anthropocentric[9] in that it implies that humanity is in a sense 'better than' (as in 'an improved version of') what came before (the exact humanist fallacy that transhumanists claim to reject), while also consequently implying that any further developments will also be improvements. It is crucial to stress that none of these claims follow from Darwin's theory of evolution and indeed fall prey to the same misunderstandings underlying Social Darwinism. Humanity is a notoriously inclusive concept. Humanity applies irrespective of the absence of certain capabilities ('differently abled') or even bodily features, so there is no reason to suspect that heightened capabilities should result in a decreased degree of humanity (let alone exclusion from it). Even more, there are no quantitative distinctions, people are not 'human, but not as much as someone else'. Heightened capabilities do not result in 'better' humans. And while some of this has historically been denied to certain groups based on skin colour or sex, it must be pointed out that there is willingness to challenge any such exclusion and accept that the views this exclusion was based on were wrong, to begin with.

So, what does it mean to be human? Is it more than just a means of claiming superiority over other life forms? Is it a purely exclusivist category of discourse and performance, as Ferrando (2014, 153) claims? There is an undeniable evolutionary case for delineating our species, irrespective of any specific quarrels over taxonomy (whether the taxon *homo sapiens* is a good choice of name; whether there might be enough reason to differentiate further within the species). This does not mean that the species *homo sapiens* exhausts the meaning of the 'human', but it is certainly a good indicator of whether an individual is a good contender for being counted among 'humans'. In that sense, belonging to our species is (currently at least) a necessary criterion for being human (and indeed legally qualifies one for claiming protection by human rights). As it is not sufficient for capturing the complexity of the discourse of humanity (which goes beyond mere biology), however, we must look further to come up with a better criterion or set of criteria that more fully represent the concept of the human. It should ideally be flexible and inclusive (in order to account for historical developments and to allow for dynamic changes to the constructed parts of the concept), but also reliable and take into account the physical foundations in some meaningful way.[10]

The solution I would like to suggest it that what more than anything else binds us together as humans is our lived experience. This probably sounds very vague initially, but this experience is partly determined by the limitations of our bodies' ways of interacting with others and with the world (our evolutionary background or phylogenesis), and partly by the historically established narrative and conceptual modes and categories of thought. In other words, it has several dimensions, among which

---

[9] Transhumanists seem to reject humanism's anthropocentrism while at the same time adopting its categories of thought, which is an ambiguous position at best.

[10] It may also be helpful to remind ourselves that the question of humanity is not one that has a right answer, but as one that is constantly re-negotiated and for which a failure of agreement does not result in a fundamental problem, but rather (at its most extreme) in a pragmatic breakdown.

the socio-historical and the embodied are the most central. Based on this notion of the human, it is possible to develop a less techno-centric, 'weak' transhumanism that is more concerned with possible human experiences than with fixing or manipulating human nature, which is a crucial shift in focus. It would consequently adopt a critical, techno-sociological perspective that is less reductive than that of current transhumanism. It furthermore allows us to place the claim that we are currently in a transhuman era in a wider context and indeed conclude that the hallmark of what transhumanists take to be the transition from human to posthuman, namely the deliberate alteration and extension of the human experience, is in fact inherently human in that is part of a practice humans have collectively engaged in for thousands of years, and indeed possibly share with at least some other species. It seems one of the hallmarks of humanity that it is constantly transforming itself as well as its environment. In that sense, transhumanism, quite ironically, seems to be the modus operandi of humanity. Whether we call this 'technogenesis' (Ferrando 2013, 28) and claim that '[t]echnology is a trait of the human outfit' (Ferrando 2013, 28), or subsume it under the broader notion of *exaptation* (Dehaene 2005) is a matter of nuance rather than principle. If we accept this view, we can only conclude that in this sense, we humans have been transhuman for a while, and there is no reason to expect this to change so fundamentally as to ever make us truly posthuman. There is therefore no threshold for where humanity ends, and the philosophical threat of 'the world's most dangerous idea' (Fukuyama 2004, 42–43) to humanity is unsubstantiated.

The logical conclusion for transhumanism as a movement, consequently, would not be to promote what is already happening on a large scale but to consider how this process can itself be changed and improved, and what the ethical, social, and political implications of deliberately structuring and directing this process are.

5.4 Human Bodies and Embodied Humans

The close connection between the human experience and the human body is central to both this narrower approach, as well as posthumanism in general. 'Humans are embodied, and so is human thought, human language, and human phenomenological reception.' (Ferrando 2014, 154) It ties in closely with embodiment theory and theories of situated cognition and is critical to collapsing the body-mind-dualism in that they are too closely interwoven to be separated. That being said, there is good reason to assume that while cognition requires a body, it is possible for a body to exist without developing or possessing cognitive capabilities. Cognition depends on certain bodily properties in a way that allows us to draw conclusions about an organism's capabilities from its neurophysiology, but also—according to theories of embodied cognition—from its physiology in general. Accurate predictions, however, can (if at all) only be made where sufficient data is available. This is partly due to the sheer complexity of brains and bodies, but also to the fact that cognition is an

5.5 Disembodied Societies

If we accept that bodily alterations by means of technology are not to be rejected in principle, as they are not inherently unethical or harmful, and if we also accept that the transhuman has in fact been humanity's modus operandi for some time now, it nonetheless poses a threat that is much more immediate; and that is a threat to society as a whole. This threat has two distinct forms, one in an increase in inequality, the other in a loss of shared experience.

5.5.1 All Posthumans Are Created Unequal

The effects of a transhumanist agenda on inequality is an area of heated debate and positions are divided along relatively clear lines between libertarian transhumanists, democratic transhumanists, and non-transhumanists. The prospect of a further increase in inequality is one that is often disputed by transhumanists, especially those of the libertarian strand, who argue that a transhumanist agenda would bring more individual freedom and that this would lead to greater equality. Non-transhumanist critics, however, argue that this is no more than a delusion, a kind of trickle-down-equality that is meant to disguise the elitist tendencies of transhumanism. Generally speaking, it has to be acknowledged that it is mainly a matter of how transhumanist policies would be implemented and sanctioned on state as well as international level, which is where the democratic transhumanists put their main focus. For what is the major distinguishing feature between eutopias and dystopias but the degree to which the distribution of technology is structured at the state level and within a society? Transhumanism, after all, is at heart a utopian endeavour, albeit one that is already having real effects.

Given that wealth distribution is one of the determining factors when it comes to life expectancy and general physical as well as mental health (Aneshensel et al. 1991; Feinstein 1993; Shaw 2005), and that the inequality in wealth distribution is rising (Davies et al. 2011; Keister & Moller 2000) due to the structural setup of our society and economic system, it seems unlikely that costly augmentations would be distributed in a fairer way that prevents a further increase in inequality. With the prospect of an even longer life-span for those who can afford it, or indeed a chance for avoiding death entirely (in any of the ways transhumanists envision, be it through changes to the body, or by transfer of memory and cognitive faculties), we also must not underestimate the role that mortality plays as 'the great leveller', a final unifying experience that affects all in the same way.[19] The prospect of those at the very top of society not dying of natural causes in the same way most people do is surely

---

[19] For a discussion of the role of death in different societies, see Faunce and Fulton (1958), or Palgi and Abramovitch (1984).

guaranteed to create civil unrest at the very least.[20] Inequalities are woven deeply
into the fabric of virtually all societies, but there is a limit to how much inequality
any particular society can cope with.

On a very practical level, our whole legal system would be in a shambles once
immortality became a reality. Not only would the severity of certain crimes be altered
dramatically, but we would also have to completely rethink the way punishment
works. If life is no longer limited in terms of duration, being sentenced to a jail
term of ten years hardly seems too daunting in the grand scheme of things. And
irrespective of the crime, a life sentence seems excessive if it means eternity. And
economic punishments would possibly mean an eternity in debt and poverty, which
is a prospect that would then again in all likelihood result in suicides, effectively
rendering it a death sentence for the less fortunate. And these are only the problems
we would be facing if immortality were available to everyone in the first place.
Luckily, however, immortality is a disputed matter even among transhumanists. But
even a structural widening of the gap in life-expectancy between different parts of
society would in effect boost many of the inequalities already in existence (Peltzman
2009).

Similarly, any alterations are motivated and implemented within societies, which
means that the underlying motivations, as well as the potential effects, are mostly
social in nature. The growing potential of the technologies concerned thus brings with
it a host of pitfalls and questions. These do, however, come in even before conscious
choices are made as to the usage of certain technologies. As Ferrando points out,

> Traditionally, scientific observations have been elaborated from a specific standpoint, which
> has been: white, Western, economically privileged, heterosexual, and male. Technology and
> science are not free from sexist, racist and Eurocentric biases; their social construction is
> embedded in their methods and practice.' (Ferrando 2014, *154–5)*

We must therefore acknowledge that even scientific discourse, down to the level
of its methodology, is fraught with bias and imbalances of power. Consequently, the
resulting technologies are inherently non-egalitarian to some degree,[21] even before
implementation and distribution come into play.

All in all, what transhumanism seems to play into is a massive power-creep. That
is to say, those who are in a privileged position will progress with greater speed and
more easily than those who are poor, sick, or disenfranchised,[22] due to the way that
alterations are likely to be made available in economically viable ways. Furthermore,
the technologies are inherently already developed in ways that privilege some parts
of society more than others, resulting in even greater privileges. While this may be an
inherent dynamic of especially western societies and globalised market capitalism,
it is clear that an acceleration in technological progress accelerates this dynamic

---

[20] Personally, I think that it is quite likely that anyone achieving such 'immortality' would likely be
killed in an uprising of the disenfranchised.

[21] For a discussion of biases contributing to failures of facial recognition technology when faced
with people of colour, see Celis and Rao (2019).

[22] This is further underlined by recent trends in growing inequality within the group of the wealthiest,
with wealth concentrated on ever fewer individuals (Perez-Arce et al. 2016).

disproportionally. It should thus come as no surprise that transhumanism often is often perceived to be an elitist endeavour, appealing only to those who are in a position so privileged that they seem to have no real other concerns.

5.5.2 Losing Common Ground

Let us now take a closer look at how the transhumanist agenda might affect human society in terms of shared (human) experience. If alterations to human bodies, their chemistry, or their ways of interacting with the world contribute greatly to an improvement in lived experience, choices will fundamentally alter an individual's experience, reducing the degree to which this experience is shared with other individuals who have chosen differently. And while I have earlier argued that human experience is fundamentally open to change and variety on a general level, it remains true—for better or for worse—that shared experience or identity contributes greatly to a society's cohesion (Smits 2003). In consequence, this means that challenges to fundamental categories along which our experience of ourselves and society is structured also pose similarly fundamental challenges to that society. It is no coincidence that societies tend to either self-regulate or self-destroy, and regulation is one of the guarantors of common frameworks.

The effects depend on the precise nature of the implemented alterations. Some glimpses of the categories that might be affected most severely are already apparent when assessing this issue within the wider conceptual framework of posthumanism. As Ferrando asks, 'will gender, race, age, and class among others, represent significant categories of reformulation? More radically, from a futuristic perspective: will posthumans need any embodiment at all?' (2014, 149) While there is indeed a sense in which many of these categories would be obsolete[23] (some of them arguably being obsolete as things stand), there is no guarantee that they would lose their importance. For one thing, people are likely to make their choices along the lines of categories that they are familiar with or that mattered to them in their choices (Jackson and Yariv 2014), meaning that they are likely to think along the lines of 'gender, race, age and class' when making choices about desirable alterations.

Furthermore, there is also a risk of simply replacing these categories with others that might be just as difficult and potentially harmful as some of the existing ones. As long as group identity remains central to our experience, people will keep defining groups by single traits. The only advantage, in this case, would be that these new categories would not already come with the same kind of historical baggage. Overall, however, the risks seem to outweigh the advantages, especially when considering how progress is being made already (albeit often too slowly) towards subverting the current categories without necessitating fundamental changes to our physical setup.

---

[23] This is of course just another way of acknowledging that many of them are socially constructed and thus contingent on historical developments, events, and choices.

5.5.3 The Many Pitfalls of Implementing Eutopias

It is especially interesting to see how the two forms of this threat are not only consequences of a radical commitment to transhumanism, but also contribute greatly to its
creation and development, revealing a structural problem. Prominent transhumanists
typically are well-situated, white, male academics or tech-entrepreneurs.[24] Similarly, there is evidence to suggest that those calling for more technological solutions
often overvalue the technology they already possess, as well as their own judgement
(Moynihan and Lavertu 2012). This is not only a tragic fact, but almost inevitable,
given that many of the issues that transhumanism is engaging with are problems that
do not rank highly among the experiences that define most people's lives (Kraus et al.
2012). Transhumanism as a movement is rooted in existing inequality and is therefore unlikely to fix it. Transhumanism, as virtually every other movement, needs
to be decolonised before it can be taken seriously as a force for positive change.
Going back to the possible definitions of the posthuman, it becomes apparent that
the individualistic framing of the posthuman, which suits its typical proponents,
seems dominant in most beyond-humanism-discourse, giving the impression of an
en vogue, technophile libertarianism. As such, it begs the questions whether there is
a strong libertarian strand to transhumanism or whether this strand is fundamentally
no more than the most up-to-date form of libertarianism.

As a social and philosophical movement, transhumanism has to answer some
uncomfortable questions about the political agenda it is pushing. The full range
from a libertarian transhumanism to a socialist or even anarchist transhumanism is
conceivable, and it is thus likely that those who comfortably identify as transhumanists at the moment are in fact more divided along traditional political lines than
they would like to think. If not, it raises questions as to how seriously some of the
promises are to be taken beyond dressing transhumanism in a more palatable guise.

There is a further problem in the inherent unpredictability of the changes that
transhumanism might bring about. The rules that govern societies' functioning are
emergent phenomena, just as societies themselves. As a result, many transhumanist
promises are speculative and premature, not as much in terms of the actual alterations
they predict to be possible in the near future, but rather in terms of the resulting
improvements in the overall experience. That is even before evaluating how much
happier people will be as a result of those improvements, and how sustainable the
changes are. People, after all, are generally bad at estimating effects from changed
material circumstances on their happiness (Frank 2005).

It is also difficult to see how some of the more visionary technologies can be
studied to a sufficient degree in an ethical way. Mind transfer, for example, is likely
to be a traumatising and disorienting experience, if not outright torture (if the issues

---

[24] One need only take a look at the makeup of the staff at the Institute for Ethics and Emerging
Technologies, the foremost transhumanist thinktank, to appreciate just how white and male
transhumanism as a movement is at its core.

related to phantom limbs and similar phenomena are anything to go by). It is impossible to test on beings incapable of offering feedback on the experience, and it is questionable at best to test it on humans before it is perfected.

In a way, however, the central flaw of transhumanist thinking is the assumption that most problems in society can be solved by means of technology. While '[i]t is commonly assumed that posthumans will lead lives and have experiences that are on the one hand unimaginable, but are on the other far superior to, i.e. much better than, anything we can experience now' (Hauskeller 2013, 2), there is not much evidence to suggest that technology is the best—let alone only—way to get there. Quite to the contrary, crucial changes to society need to precede any structural and organised implementations of the majority of transhumanist policies, if there is to be any chance of an actual increase in equality and thus in freedom. Immortality and expanding the human capacity to experience and interact with the world are reasonable goals, but only once problems such as inequality, starvation, and the limitations of resources (including but not limited to energy and space) have been solved.

5.5.4 The Voyage Home

Reading about the issues that concern transhumanists often feels like a funding pitch of a hip start-up with great marketability, which has not really thought through its actual product. Most of transhumanism's promises, should they hold true, will surely feel like pyrrhic victories, with the reality of things far more sobering than imagined. Much like in one of the Star Trek franchise's most popular films, we might one day find ourselves in a situation where we realise that it would have been better to sort out the actual problems at the time rather than developing technological means of glossing over them. If the posthuman is to have moved beyond inequality, racism, and sexism, it is surely better to confront these issues rather than fundamentally change who we are and hope that our problems will go away as a result. Transhumanism's current agenda gives the uneasy impression that some suspect that prejudice and inequality are justified by our physical existence after all, and that we cannot hope to eliminate them without eliminating this justification.

As a result, transhumanism is a movement for progress rather than a progressive movement, fixated on technological solutions whose promises distract from the need for social change in a society that is technologically already capable of fixing many of its fundamental problems, but unwilling to commit to the necessary effort. In such a society, the availability of further enhancing technology cannot but deepen divisions and promote inequality. While we have all been transhuman for some time, transhumanism (with the exception of some of its more down-to-earth goals) largely remains a project for the elite. This elite is more preoccupied with finding new ways of expanding their experience of a somewhat over-saturated life that is exhausting the consumption-limits of our bodies, than it is with changing the general setup of our society in ways that would enable more fundamental freedom for everyone, irrespective of social standing and wealth.

None of this is to say that there are no legitimate or constructive suggestions put forward by transhumanists, or that all transhumanists are egocentric elitists. Instead, it is the observation that in order for transhumanism to grow into a properly progressive movement and to gain more traction within wider society and political manifestos, it needs to realign its priorities. As it stands, it rightly presents itself as a movement that has its main proponents amongst philosophers and tech-gurus, people who are occupied with questions of what it means to be human and how this experience can be enhanced. And while these are honourable and incredibly interesting pursuits, most people would probably rather consider these issues with a full belly and a positive account balance.

References

Aneshensel, C.S., et al. 1991. Social structure, stress, and mental health: Competing conceptual and analytic models. *American Sociological Review* 56 (2): 166–178.

Bostrom, N. 2003. Human genetic enhancements: a transhumanist perspective. *The Journal of Value Inquiry* 37 (4): 493–506.

Bostrom, N. 2001. Transhumanist Values. Version of April 18, 2001. https://nickbostrom.com/tra/values. Accessed 12 Aug 2022.

Braidotti, R. 1996. Signs of wonder and traces of doubt—On teratology and embodied differences. In *Between monsters, goddesses and cyborgs—Feminist confrontations with science, medicine and cyberspace*, ed. N. Lykke and R. Braidotti, 135–152. London: Zed Books.

Bunge, M. 1980. *The mind-body problem: A psychobiological approach.* New York: Pergamon.

Celis, D., and R. Meghana. 2019. Learning facial recognition biases through VAE latent representations. In *Proceedings of the 1st international workshop on fairness, accountability, and transparency in multimedia (FAT/MM '19), association for computing machinery*, 26–32. New York.

Crawford, C.S. 2014. Phantoms in the mind: The psychogenic origins of ethereal appendages. In *Phantom limb: Amputation, embodiment, and prosthetic technology*, 107–148. New York: NYU Press.

Davies, J. B., et al. 2011. The level and distribution of global household wealth. *The Economic Journal* 121 (551): 223–254.

Dehaene, S. 2005. Evolution of human cortical circuits for reading and arithmetic: The 'neuronal recycling' hypothesis. In *From monkey brain to human brain*, ed. S. Dehaene, J.R. Duhamel, M. Hauser, and G. Rizzolatti, 133–157. Cambridge, MA: MIT Press.

Dennett, D. C. 1991. *Consciousness explained*. Little, Brown, and Co, Boston.

Dennett, D. C. 2018. Facing up to the hard question of consciousness. *Philosophical Transactions of the Royal Society of London. Series B, Biological Sciences* 373 (1755): 20170342.

Faunce, W. A., and L. F. Robert. 1958. The Sociology of death: A neglected area of research. *Social Forces* 36 (3): 205–209.

Feinstein, J. S. 1993. The relationship between socioeconomic status and health: A review of the literature. *The Milbank Quarterly* 71 (2): 279–322.

Ferrando, F. 2013. Posthumanism, transhumanism, antihumanism, metahumanism, and new materialisms: Differences and relations. *Existenz* 8 (2): 26–32.

Ferrando, F. 2014. The Body. In *Post- and transhumanism: An introduction*, vol. 1, ed. R. Ranisch and S.L. Sorgner. Beyond humanism: Trans- and posthumanism, 149–162. Frankfurt am Main: Peter Lang Publisher.

Frank, R. H. 2005. Does money buy happiness. *The Science of Well-Being* 461–473

Fukuyama, F. 2004. The world's most dangerous ideas. *Transhumanism. Foreign Policy* 144: 42–42.

Hauskeller, M. 2012. Reinventing Cockaigne: Utopian themes in transhumanist thought. *Hastings Center Report* 42 (2): 39–47.

Hauskeller, M. 2013. *Utopia in trans-and posthumanism.* Frankfurt am Main: Peter Lang Publisher.

Hughes, J. 2012. The politics of transhumanism and the techno-millennial imagination, 1626–2030. *Zygon* 47 (4): 757–776.

Jackson, M.O., and L. Yariv. 2014. Present bias and collective dynamic choice in the lab. *The American Economic Review* 104 (12): 4184–4204.

Kannengiesser, U., and J. S. Gero. 2012. A process framework of affordances in design. *Design Issues* 28 (1): 50–62.

Keister, L. A., and S. Moller. 2000. Wealth inequality in the United States. *Annual Review of Sociology* 26: 63–81.

Kim, J. 1984. Concepts of supervenience. *Philosophy and Phenomenological Research* 45 (2): 153–176.

Kraus, M. W., et al. 2012. Social class, solipsism, and contextualism: How the rich are different from the poor. *Psychological Review* 119 (3): 546.

Kurzweil, R. 2005. *The singularity is near—When humans transcend biology.* New York: Viking.

Moynihan, D. P., and S. Lavertu. 2012. Cognitive biases in governing: Technology preferences in election administration. *Public Administration Review* 72 (1): 68–77.

Palgi, P., and H. Abramovitch. 1984. Death: A cross-cultural perspective. *Annual Review of Anthropology* 13: 385–417.

Peltzman, S. 2009. Mortality inequality. *The Journal of Economic Perspectives* 23 (4): 175–190.

Perez-Arce, F., et al. 2016. Trends in Inequality. In *Inequality and opportunity: The relationship between income inequality and intergenerational transmission of income,* 9–22. RAND Corporation, Santa Monica

Shaw, J. W., et al. 2005. The determinants of life expectancy: An analysis of the OECD health data. *Southern Economic Journal* 71 (4): 768–783.

Smits, K. 2003. Liberalism's identity problem. *Polity* 35 (3): 347–367.

Sorgner, S. L. 2016. Transhumanismus: die gefährlichste idee der Welt!?. Herder, Freiburg.

Wilmer, H. H., L. E. Sherman, and J. M. Chein. 2017. Smartphones and cognition: A review of research exploring the links between mobile technology habits and cognitive functioning. *Frontiers in Psychology*, 8: 605.

Robin Markus Auer is currently working towards a Ph.D. as part of an interdisciplinary research project on automated creativity in literature and music at TU Braunschweig, funded by the federal state of Lower Saxony, Germany. His work focuses on the interplay between human and machine creativity in coupled embodied creative systems, and how this interplay reflects on and questions traditional assumptions about (particularly artistic) creativity and art itself. Previously, he has completed two Master's degrees (English and Philosophy) at Merton College (Oxford) and Heidelberg University. His Master's thesis in philosophy developed a 'Situated Semiosis Theory of Consciousness' which seeks to explain human 'phenomenal' consciousness (in the context of Chalmers' 'hard problem of consciousness') as a result of increased complexity within an embodied and embedded dynamic semiotic system. His research interests include theories of consciousness and creativity, philosophy of mind, science and language, as well as semiotics, and literary theory.

Chapter 6
Evolving the Natural-Born Cyborg

Belinda Marshall

Abstract Transhumanism is the concept of eliminating the weaknesses and "failings" of our evolved biology, often through the use of technological artefacts. Without such limitations as aging, physical strength, capacity for knowledge retention, and so on, Transhumanists argue that we can "evolve into something better". In his book *Natural-Born Cyborgs* (2003), Andy Clark claims that we need not fear the notion of the 'cyborg'—as we ourselves are already naturally predisposed to a cyborg-esque nature. Drawing upon works within cognitive science that tackle extended cognition, Clark demonstrates that humans have a natural instinct to incorporate tools due to the plasticity of our brains, and our already well-demonstrated capacity to integrate technology in such a way that it becomes an extension of self. Clark's notion of the natural-born cyborg creates an interesting point of discussion, ultimately leading us to the question of whether Transhumanism is the inevitable next step for the human race. At first consideration, it would appear that would be a reasonable assumption—however, turning to the field of extended cognition, I believe we can find interesting phenomenological links between the importance of embodiment and presence in the world, and how we can retain such facets of humanity, whilst still taking advantage of our natural abilities to create and adapt to technology and technological items. The conclusion of such, leads to Virtual Reality technology; a relatively new form of technology, which I argue holds qualities which set it apart from other forms of technology—qualities which allow the advancement towards "posthuman", whilst countering some of the potential downsides of human-technological integration.

6.1 Transhumanism

Transhumanism is a philosophy which began a cultural movement, which thus sparked further study and debate, with particular focus on seeking *"the continued evolution of human life beyond its current human form as a result of science and*

B. Marshall (✉)
University of St. Andrews, St. Andrews, UK
e-mail: Bjr4@st-andrews.ac.uk

technology guided by life-promoting principles and values" (*The Transhumanist Reader*, p. 1), and is most well known as being conceived by philosophers Max More and Natasha Vita-More. There are several interpretations and perspectives within the Transhumanist ideology, but in his edited volume *The Transhumanist Reader* (2013), Max More aims to summarise the central themes, values, and interests of the Transhumanist, giving three examples of the definition, which display coherent and unanimous agreement on the core principles:

> According to my early definition (More, 1990), the term refers to:
>
> *Philosophies of life (such as extropian perspectives) that seek the continuation and acceleration of the evolution of intelligent life beyond its currently human form and human limitations by means of science and technology, guided by life-promoting principles and values.*
>
> According to the Transhumanist FAQ (Various, 2003), transhumanism is:
>
> *The intellectual and cultural movement that affirms the possibility and desirability of fundamentally improving the human condition through applied reason, especially by developing and making widely available technologies to eliminate aging and to greatly enhance human intellectual, physical, and psychological capacities.*
>
> A corollary definition (also from the FAQ) focuses on the activity rather than the content of transhumanism:
>
> *The study of the ramifications, promises, and potential dangers of technologies that will enable us to overcome fundamental human limitations, and the related study of the ethical matters of developing and using such technologies.* (The Transhumanist Reader, p. 4)

Essentially, by applying science and technology to the current state of the natural human being, the Transhumanist aims to overcome the limitations imposed by our biological and genetic heritage; human nature is not "an end in itself", but just one point along the evolutionary pathway—we ought to reshape our own nature in ways which seem both desirable and valuable, and become *"something no longer accurately described as human - we can become posthuman."* (The Transhumanist Reader, p. 4) This entails eradicating disease and even the process of aging; increasing our physical capacities and freedom of form; enhancing our cognitive abilities and refining our emotional states—but without any sense of an "end goal", just the continuous attempt to improve and evolve, to advance the human race alongside the advancements of technology. In *Natural-Born Cyborgs*, Clark appears to create a functional foundation for the potential "ideal" Transhumanist future:

> *Such technologies will be less like tools and more like part of the mental apparatus of the person. They will remain tools only in the thin and ultimately paradoxical sense in which my own unconsciously operating neural structures (my hippocampus, my posterior parietal cortex) are tools.* (Clark, NBC, p. 7)

6.2   The Natural-Born Cyborg

When considering the ideological concept of Transhumanism, it is interesting to also consider the concept of the 'Natural-Born Cyborg', as posed by Andy Clark.

Natural-Born Cyborgs was written in 2003, one of Clark's many contributions within the fields of cognitive science, philosophy of mind, and philosophy of technology. The term "cyborg"—a contraction of "cybernetic organism"—aims to describe a being, which is both biological and organic, but also incorporates technological components which—theoretically—improve human abilities, or restore absent human functions. Although such cyborg-esque beings have been portrayed in the media since the early twentieth century, particularly within the genre of science fiction, the first use of the term was by Manfred Clynes and Nathan S. Kline in *Cyborgs and Space* (1960):

> *For the exogenously extended organization complex functioning as an integrated homeostatic system unconsciously, we propose the term "Cyborg." The Cyborg deliberately incorporates exogenous components extending the self-regulatory control function of the organism in order to adapt it to new environments. If man in space, in addition to flying his vehicle, must continuously be checking on things and making adjustments merely in order to keep himself alive, he becomes a slave to the machine. The purpose of the Cyborg, as well as his own homeostatic systems, is to provide an organizational system in which such robot-like problems are taken care of automatically and unconsciously, leaving man free to explore, to create, to think, and to feel.* (Cyborgs and Space, p. 27)

The notion of cyborgs itself, however, simply provides a definition of this specific type of human and technological integration; what Clark aims to argue in *Natural-Born Cyborgs*, is that we are all *already* cyborgs by nature. Such a concept has previously been explored, particularly in postmodernist literature (see *A Cyborg Manifesto*—Donna Harraway), but what struck Clark, was the notion that this was the literal and scientific truth:

> *The human mind, if it is to be the physical organ of human reason, simply cannot be seen as bounded and restricted by the biological skinbag. In fact, it has never been thus restricted and bound, at least not since the first meaningful words were uttered on some ancestral plain. But this ancient seepage has been gathering momentum with the advent of texts, PCs, coevolving software agents, and user-adaptive home and office devices. The mind is just less and less in the head.* (Natural-Born Cyborgs, p. 4)

In a continuation of previous works on extended cognition (see *The Extended Mind*—Clark and Chalmers 1998), Clark aims to reassess the traditional notion of the cyborg—seeing it not as some futuristic cultural icon of "human–machine hybrids", but revealing it as an oddly concealed vision of our own biological nature. For Clark, what is important to note about the human brain and intelligence, is its unique capacity for complex relationships with non-biological constructs, props, and aids—which does not necessarily require the use of physical wire and implant mergers, as much as it relies on our openness to mergers of information processing. Without the silicone intrusion as often depicted by the notion of the cyborg, we can still recognise the integration of human and technology—for example, thinking via the act of writing; such external props and aids to our problem-solving systems are deeply integral to our everyday activities, which leads Clark to argue that these props should be accepted as part of human intelligence, in the sense that "*[s]uch tools are best conceived as proper parts of the computational apparatus that constitutes our minds.*" (p. 6, NBC).

6.3   Crossing the Line Between Natural-Born Cyborgs and Transhumanism

Although there is no one particular definition of a Transhumanist, nor any direct distinction between the Transhumanist's notion of the ideal "post-human" and current conceptions of the "cyborg", it appears that through the rough definitions of both, we can draw direct parallels between the two. The cyborg—the cybernetic organism— is a technologically-enhanced human; a technologically-enhanced human race is the goal of Transhumanism. If we take Clark's conclusion to be true—that our brains are biologically geared towards utilising technology for improved performance in a variety of functional and practical ways—we may consider that, given the ever-increasing growth of technological invention and use, we are inevitably going to become intertwined with technology in ways, which *do* fit the Transhumanist ideology. From this conclusion, however, a few issues arise. Firstly—although the cyborg, according to its original conception and definition, aims to integrate technology to outsource certain processes and tasks *"leaving man free to explore, to create, to think, and to feel"* (Cyborgs and Space, p. 27), Transhumanism, seems to imply a more "invasive" advancement of humanity, with no clear end-goal, or point of limitation. In this section, I hope to clarify why some may consider fully-fledged Transhumanism to be problematic, by exploring the shift from using technology as an extension of self—as a tool for human use, to the Transhumanist goal of using technology to evolve the human into something new entirely.

The distinction is minor, yet crucial—in attempting to use technology to outsource minor cognitive operations, as is possible via the concept of the 'natural-born cyborg', we allow ourselves to place a more active focus on other, more useful, productive, or fulfilling human pursuits; we are placing technology in service to ourselves as human beings, treating it as an *extension* of the self, without directly attempting to create a fundamental shift in what it is to "be human". In the case of Transhumanism, however, by striving to use technology to continuously enhance and improve our natural physical and mental state with no clear end-point or goal, the Transhumanist appears almost indifferent to the facets of "being human" that ought *not* to be technologically enhanced—the things which are more beneficial or appreciated in their current form, whether capable of "improvement" or otherwise. Such a criticism of Transhumanism is undeniably vague; is there any true method of knowing which part of our natural and biological way of being is crucial to us? Or even, any way of knowing when we've "gone too far"? One may already argue that our lives consist of "too much technology"—many *do* argue that as the presence of technology in our everyday lives increases, that we have become more distracted, less social, and so on—however, there is no certain account of whether sacrificing attention span for productivity, for example, is the "right" thing to do. Regardless, it would be fair to assume that the majority of human beings would prefer to retain some of their current form of self—either alongside, or instead of, technological enhancement.

If we are—as Clark argues—natural-born cyborgs, and thus inevitably hurtling towards the Transhumanist ideals due to our natural disposition towards technological integration, the important question is no longer "should we?", but "what is the best possible outcome of this and how can we achieve it?" I believe the answer to this lies in virtual reality technologies—however, before unpacking this, it is important to first clarify some of the worrying facets of the Transhumanist ideology.

In his 2017 book *To Be A Machine*, Mark O'Connell touches upon Transhumanism, and some of the ways in which it cannot fully account for the human experience:

> *Transhumanism is a liberation movement advocating nothing less than a total emancipation from biology itself. There is another way of seeing this, an equal and opposite interpretation, which is that this apparent liberation would in reality be nothing less than a final and total enslavement to technology.* (To Be A Machine, p. 6)

In this travelogue-style book, O'Connell seeks to research various Transhumanist positions and goals, through journeys to various lectures, cryogenic labs, and so on—ultimately written through the lens of a man, new to fatherhood, aiming to pinpoint what exactly it is that makes us human, and why exactly Transhumanists are attempting to use science and technology to "transcend" it. A particularly poignant passage in the book is written following a meeting with neuroscientist Randal Koene to discuss his work on mind uploading:

> *I lay there, idly considering hauling myself out of bed to join my wife and my son, who were in his bedroom next door enjoying a raucous game of Buckaroo. I realised that these conditions (head cold, hangover) had imposed upon me a regime of mild bodily estrangement. As often happens when I am feeling under the weather, I had a sense of myself as an irreducibly biological thing, an assemblage of flesh and blood and gristle. (...) I was aware of my substrate, in short, because my substrate felt like shit. And I was gripped by a sudden curiosity as to what, precisely, that substrate consisted of- as to what I myself happened, technically speaking, to be. (...) It turned out that I was 65 percent oxygen- which is to say that I was mostly air, mostly nothing. After that, I was composed of diminishing quantities of carbon and hydrogen, of calcium and sulfur and chlorine, and so on down the elemental table. I was also mildly surprised to learn that, like the iPhone I was extracting this information from, I also contained trace elements of copper and iron, and silicon. What a piece of work is a man, I thought; what a quintessence of dust. Some minutes later, my wife entered the bedroom on her hands and knees, our son on her back, gripping the collar of her shirt tight in his little fists. She was making clip-clop noises as she crawled forward, and he was laughing giddily, and shouting "Don't buck! Don't buck!". With a loud neighing sound, she arched her back and sent him tumbling gently into a row of shoes by the wall, and he screamed in delighted outrage, before climbing up again. None of this, I felt, could be rendered in code. None of this, I felt, could be run on any other substrate. Their beauty was bodily, in the most profound sense, in the saddest and most wonderful sense. I never loved my wife and our little boy more, I realised, than when I thought of them as mammals. I dragged myself, my animal body, out of bed to join them.* (To Be a Machine, pp. 68–69)

For O'Connell, his journey of research into Transhumanism, meeting many of the famous contributors to the field, interacting with various fields of Transhumanist research, did not leave him compelled towards their goals. Despite accepting that perhaps in *some* sense, we are *already* living up to Transhumanist ideals— O'Connell's research shines a light on the side of Transhumanism which doesn't seem to account for the many rich experiences we have as human beings—that could only be as they are, despite our biological and physical limitations (as they are often seen as being by Transhumanist thinkers).

In *Natural Born-Cyborgs*, Clark also dedicates an entire chapter (*Bad Borgs?*, p. 167) to the potential "*closures, dangers, invasions, and constraints*" of these "*hybrid dreams*" (NBC, p. 167)—which although are more specifically directed towards his own conception of the natural-born cyborg rather than directly to Transhumanist ideology, do directly correlate with the potential problems of a Transhumanist future, given the direct links (and potential shift) between the two. Clark's examples include, for example, the intrusive nature of technology—which is particularly worrisome as we become more integrated with it, especially if we consider the lack of security and privacy which exists within external pieces of technology, and imagine similar breaches to technologies which are either closely coupled with our cognitive systems, or directly implanted into our physical bodies. Another of Clark's examples includes the potential of "overload"—much in the way that our email inboxes can become quickly filled with what he calls "e-stodge"—emails which are "filling without being necessary or nourishing" (NBC, p. 176)—our lives run the risk of being filled with information and communication that is simply more than we need, or can comfortably handle. As suggested by Neil Gershenfeld, the root cause of such, is the "*deep but unnoticed shift in the relative costs, in terms of time and effort, of* generating *messages and of* reading *them.*" (Kelly 1994), thus, as our ease and ability to output information to others is increased, our likelihood of *receiving* more information increases also. The answer to this problem, Clark states, is "simply to unplug"—although he further admits that this isn't always a possibility and that perhaps some kind of intelligent filtering software, or some form of "etiquette" which "*reflects the new cost/benefit ratio according to which the receiver is usually paying the heaviest price in the exchange.*" (NBC, p. 177)—although, fast-forward seventeen years since the publication of this book, and we can comfortably ascertain that no such solution has been found or widely implemented. Clark touches upon several other potential dangers and constraints within this chapter, however, I will address a few of these in greater detail when introducing the potential benefits of using Virtual Reality as an alternative to Transhumanism.

Considering both O'Connell's more phenomenological view on how Transhumanist ideals have the capacity to diminish or undervalue what it is to "be human", alongside Clark's more practical approach to the potential pitfalls of technological integration, it is clear to see already that Transhumanism may not be an ideal goal for humanity. Whether we intuitively consider Transhumanism to be something distinctly un-human and undesirable, or simply consider that there are too many potential logistical and technological issues which may arise from becoming so intertwined with technology that we essentially evolve into an entirely new type of being, we can readily come to the conclusion that perhaps an alternative approach to our biological predisposition to integrate with technology should be explored.

6.4   Computer-Mediated Reality and the Impact of Modern Technology

Virtual reality is a relatively new form of technology, one which aims—via the use of a variety of forms of technological equipment—to present the user's senses with made-up information or data, in an attempt to alter their perception of reality, or immerse them in an "alternate" reality—which is often a three-dimensional, computer-generated environment that can be explored and interacted with. Although "virtual reality" takes a variety of forms, it is always implemented through the use of some kind of computer technology—be that through a traditional desktop or laptop computer, more portable smartphones, or more sensory-immersive computational technology such as the archetypal headgear or goggles, omnidirectional treadmills, glove-style controllers, and so on. Full or partial sensory immersion is often mistaken as the quintessential form of virtual reality—however, virtual environments can be—and are most commonly—accessed via a simple keyboard and mouse, or a games console controller. Virtual reality technology simply requires synchronicity between user, hardware, software, and (sometimes) sensory technology to create a 'sense of presence' as it were, in order to fully immerse the user in the particular virtual environment of choice; a reduction in physiological immersion does not necessarily lead to a reduction in cognitive immersion—in fact, one could argue that given the limitations of the newly-designed virtual reality technologies, which rarely work seamlessly with our physiology, the more "simplistic" forms harness a larger capacity for immersion within the virtual environment. One example of this is "cyber-sickness"—a motion-sickness-esque illness which arises when information presented to the user's eyes and ears is in conflict—the discomfort of such, reduces the user's capacity for immersion and virtual presence, as they are distracted by their real-world ailment. Virtual reality has several real-world uses and applications, alongside its growing prevalence in the gaming industry, it allows us to complete tasks which are either dangerous, impractical, or costly in real life, taking virtual risks to provide real-world experience. Examples of such can include trainee pilots or surgeons, virtual heritage tours or phobia treatments, and so on.

Virtual reality technology is a form of computer-mediated reality technology, a subset of modern technology which attempts to blur the boundaries between technological reality, and actual primary reality; another common example of such, is augmented reality technology, which is a form of technology which aims to augment primary reality by placing technological images and objects into it. This could include something as simple as a satellite navigation system in a car, displaying a digital version of the driver and road to aid navigation, or even a holographic projection used in a concert. Augmented reality technology, much like virtual reality technology, is also increasingly popular in the entertainment industry—particularly for gaming, demonstrated by a recent increase in mobile games, which take advantage of a smartphone's camera application to display the real life scenery through the screen, but with additional digital imagery. *Pokemon Go* was a huge example of popular augmented reality technology, as a mobile game that allowed the users to explore the real-world environment to find "pokemon"—small digital monsters which the players could find, collect, and trade, similar to its previous *virtual* counterpart, a video game in which the purpose of the game was the same as its AR addition, but in an entirely virtual environment. It is important, I believe, to understand the crucial differences between augmented reality and other computer-mediated reality technologies, and virtual reality technologies, as there are a few elements of virtual reality technology that are unique to itself.

The first and most crucial difference between augmented reality and virtual reality is that where augmented reality aims to place the digital artefacts into primary reality, virtual reality aims to remove the user's sense of primary presence and immerse them into a digital environment. Where AR allows the technological to enter our world, VR tempts us into the technological world, leaving primary reality behind. This is particularly crucial when we consider some of the issues raised by an ideal Transhumanist future, or the complications of our cyborg-esque self entering an era of increasing technological advancement. For example, looking back to Clark's discussion on the potential "overload" posed by increased use of technology—augmented reality technology increases the amount of technology *in* our everyday lives. As more technology is embedded in the primary reality in which we are bound, as it becomes increasingly realistic—or even, as is the case with the satellite navigation technology, increasingly necessary and relied upon—we are evermore likely to feel the weight of excessive technological input. Such is the case with most other forms of technology—computer-mediated reality or otherwise—we are rarely without a smartphone to hand, the chance of an attention-grabbing phone call or email at any moment; it is almost mandatory to own a computer or at least be able to access the internet, even for basic (yet necessary) everyday tasks such as banking, gaining employment, and so on; education settings are highly reliant on a variety of forms of technology for both teaching and learning; the list goes on. Modern computing technology is almost inescapable in the present day Western world, which has already resulted in an entirely modern set of problems—which I believe is an important thing to take into account when considering the Transhumanist desire to increase our use of, and integration with, technology.

Research shows that technology changes the way that we think (Carr 2008; Richtel 2010); in order to operate a computer, for example, one must hold a basic understanding of the way the interface of a computer operates, and interact with it accordingly—you cannot simply ask your computer to open a particular application, but must learn to navigate the linear interface—which is problematic as interaction with technology is increasingly frequent, and evidently having an impact on the way that we feel and behave within society (Greenfield 2003). Nicholas Carr, throughout several books on technology and its impact on human experience, explores how technology can rewire our brain's physiology (The Shallows 2010a), as well as how the increase in use of technology is strengthening our ability to scan information efficiently—at the cost of no longer being able to stay focused and imaginative, as reading books allows (Carr 2010b). With particular emphasis on the difference between the impact of technology, or reading books, on attention span, Carr uses a metaphor outlining the difference between scuba diving and jet skiing; scuba diving, like book reading, submerges the diver into a slow-paced, quiet setting with few distractions, and limited visual stimulation. As a result of this limited information, a high level of focus is required, as well as clarity of thought. Jet skiing, on the other hand, can be likened to using the internet—the skier is travelling at high speed above the water's surface, with several visual distractions, each of which can only be afforded fleeting attention. The increase in use of technology, and the internet, in particular, poses a threat to our natural state of being, especially as new generations are growing up with frequent exposure to technology as the norm (Taylor 2012). Carr's research is particularly interesting for two reasons: firstly, that in criticising the impact of technology on the way we think and act, he uses reading as a comparison. For Carr, as shown above, reading enhances our cognitive ability, whereas the use of technology inhibits—or at least, changes it. An interesting facet of Virtual Reality technology, especially in the form of video games and virtual environments, is that it can arguably be likened to the experience of reading a book. Rather than mindlessly flitting between pieces of information whilst browsing the internet or using social media, exploring virtual environments can indeed be very similar to the experience of the scuba diver; the pace of virtual environments can be slow, there is capacity for exploration and focus; video games often include a narrative, allowing the player to immerse themselves in a story, much as a reader would immerse themselves in a book—although, of course, there is a clear distinction in that the necessity for imagination is mostly diminished. Research actually shows, that despite the many studies (as cited above) which show the negative impact of technology on ourselves, our brains, and our lived experience—playing video games can actually enhance several skills, "including low-level vision (enhanced contrast sensitivity function), various aspects of attention (ability to monitor several objects at once, to search through a cluttered scene, to detect an event of interest in fast-forwarding video), more complex task constructs (multi-tasking, task-switching) and, finally, a general speeding of perceptual processing." (Bavelier 2012). This is just one example of how Virtual Reality technologies can differ from other forms of modern technology, although I will further explore the benefits of Virtual Reality in the next section.

The above research is interesting, particularly as it demonstrates not only that we are integrating with technology in such a way that it is *already* altering our biologically human state, but that these alterations are in fact quite detrimental—and although the Transhumanist may rebut this by arguing to design technologies which specifically counteract such problems, it still seems likely that those new and different technologies may simply cause new and different problems. If we accept that technology is having a negative impact on human beings, surely the only solution to this, aside from the extremely unlikely occurrence of not using it, is creating technologies which are *less* technological—technologies which complement our human experience. The above is just one of many issues which arise from modern technological use, and whilst I can't go into much more here, it is worth noting that other extensive problems include a profound impact on our lived experience (see Wallsten 2013), that technology makes us less sociable (see Waytz and Gray 2018), and that it is even addictive to the point of causing great disruption to our daily lives (see Young 1998; Salehan 2013).

6.5 Virtual Reality as an Alternative to Transhumanism

Above, I touched upon one example of how Virtual Reality technology can be distinguished from other forms of modern technology, thus potentially offering a solution to the way we currently use technology and the problems, which follow. However, referring again directly to the potential problems of Transhumanism (as discussed above), I would like to further argue the case that Virtual Reality could be used as an alternative to Transhumanist goals. Whilst Virtual Reality is still capable of expanding our capabilities and can be used as a tool to diminish some of the limitations of being human, as is desired by Transhumanists, it doesn't fundamentally change what it is to be human (or at least, no more so than the standard technologies we are using today), and isn't as susceptible to the problems which arise from the concept of being so heavily integrated with technology.

Referring back to the case of "overload" mentioned above, as discussed by Clark in his chapter 'Bad Borgs?'—once the user has become fully immersed in the virtual environment of choice—regardless of whether that's a simple video game or a full sensory-immersive environment—they have full control over how much engagement they have *with* that environment. Many modern games simulate real world tasks and activities, often including monotonous and repetitive actions to partake in and complete, social interactions (be that with other real human beings, or artificial interlocutors), travel and exploration, and so on; although similar real world tasks (having to travel to a destination, having to socialise, having to complete tasks to achieve a goal, and so on) may cause fatigue and overload in the real world—often *especially* due to the increase in technological intrusion—such tasks in virtual worlds are often seen as enjoyable. People actively make the decision to enter these virtual environments and complete said tasks—oddly enough, even when they aren't necessarily "enjoying" the task at hand—yet it is rare that the risk of overload is present.[1] The crucial difference here is, to mirror Clark's sentiments towards excessive "e-stodge", that one can simply "turn off"—one *can* escape a virtual environment at any given moment, be that to enter a different virtual environment, or literally come "back to reality". It appears that whereas increased integration with technology runs the risk of overwhelming us if we were to focus on more forms of technology that we could willingly enter and escape, that we had greater control over, the risk of such would be greatly diminished.

A real world example of how this could practically be implemented is in workspaces; it is a common stereotype of the modern day that we become stuck in "work-mode" at work, that we come home to spend time with our families and relax, but instead find ourselves still opening our email inboxes to check work-related emails; our minds will be half-present, half-still in the office. The technology used at work is, commonly, technology accessed at home; we can still access the internet, most of us will still have our smartphones to hand—increasing the likelihood of being contacted about work-related issues (or even making the contact ourselves), often it

---

[1] For more on the topic of video games and their similarity to real world tasks, see Chalk and Powell (2020), Kennett (2015).

is possible to remotely access particular work-related software or web applications. Although we physically travel to and from a set location, in which we spend our eight or so paid hours completing tasks for financial gain, and then physically travel back home—we seemingly don't leave that much behind at all, beyond the physical environment itself. Imagine instead, in some technologically-advanced future, workspaces which *if possible* to be made virtual—had been. Rather than commuting to the office, one would sit at a desk with the necessary equipment, log into a virtual environment, and complete all necessary tasks as prior. Perhaps the user is wearing sensory headgear which simulates them being sat at a desk, next to other workers on their virtual desks; perhaps the interface is a very basic set of symbols allowing the worker to access the necessary applications to fulfil their tasks—perhaps it's a full attempt at digitally replicating a physical workspace. The crucial point is that the place of work—the part of life, which has demanded the most increased use of, and integration with, technology—is no longer a part of the worker's primary reality. At the end of their workday, they can sign out of their virtual computer, virtually say goodbye to their colleagues, and then take off their virtual reality headgear (or whichever particular form of technological equipment is necessary per situation)— finding themselves back in a primary reality in which they are fully removed from their workspace. In some manner, it appears that in moving the most at risk of causing overload technologies into an escapable environment, the overall negative impact of technological integration could be decreased.[2]

It isn't just the solution to the problem of overload which makes virtual reality technology such a potentially positive form of technology either; in fact, it seems exempt from, or at least ill-fitted to, almost every potential criticism of technological integration offered by Clark in *Natural-Born Cyborgs*. Clark poses the issue of inequality—the reduced access to technology in lesser developed countries, along-side advances in human capability offered *by* technology in more developed countries, leading to greater disparity of equal opportunities or abilities. Although this *does* apply to virtual reality technology in the sense that it is still—at present— too costly for the average household, nevermind for mass distribution, it does offer solutions to some problems of inequality raised by technology. For example, *within* virtual environments, factors such as wealth, appearance, employment status, and so on, are rarely visible or applicable. Two human players in the same virtual environment, completing the same virtual tasks, could own the same virtual items, have had the same virtual achievements—yet have vastly different achievements and qualities in real life. Where economic, cultural, and other real-life disadvantages cause issues in real-life environments, they are essentially invisible in virtual environments. Skills gained in virtual worlds through the use of virtual artefacts, can—for the most part—be equally acquired; integration with technology *within* virtuality, allows for equal gains from such integration. Similarly with Clark's discussion on the potential danger of *alienation*, whereby technology disconnects us from reality as we engage with it more—our social habits are changed to suit the technologies around us, we

---

[2] The concept of virtual workplaces and their potential benefits and feasibility is further explored by Virtual Reality researcher Michael Heim in his book Virtual Realism (1998).

are increasingly concerned with interacting with efficient virtual versions of things rather than maintaining contact with less-efficient human beings. This, again, seems to barely apply to virtual reality technologies—even seemingly offering a solution. Virtual environments are often (although not always) social environments; although some, such as single-player video games, immerse the user in a virtual environment with only artificial interlocutors to interact with, many will allow a large amount of users to play in the same virtual environment—be that to work as a team towards a goal, work against each other in some competitive manner, or simply act individually with the option of socialising or teaming up. Although increased interaction with virtual environments is technically still an alienation from primary reality, there seems to be no sufficient argument as to why socialising and communicating in a virtual format is any less fulfilling or important. If anything—as Clark also points out in *Natural-Born Cyborgs*—virtual spaces actually increase opportunity to find and interact with enthusiasts of similar hobbies or interests to oneself; what may be a niche and unique hobby in the context of one's geographical community, may be a common interest in one's virtual community—perhaps an immersive engagement with virtual communities may actually lead to greater achievements, as one is further encouraged to engage with things that they are passionate about, potentially share experiences and knowledge with others, and even widen their social network in *both* the virtual and primary realities.

Referring back to O'Connell's quote from *To Be a Machine*, in which he speaks of his wife and child playing, in a scene which simply couldn't be "rendered in code", the situation is a little more complicated. In one sense, O'Connell is almost certainly correct—it is genuinely difficult to make the claim that any virtual imitation of such a scenario would be anything but lacking; the human element of the author's struggle to overcome his physical ailments, intermingled with his human desire to engage in play with his family, the physical bodily elements of such—all of these elements simply *are* human. However, I don't think it is as black and white as it initially seems. As discussed above, I believe it would be difficult to genuinely argue that meaningful communication and experiences cannot happen virtually; there are already thousands of unique blogs, social media pages, online videos, and so on, which so clearly detail and describe rich and unique virtual experiences—both alone and with others—which fulfill social desires, allow for the element of surprise, joy, "togetherness".[3] Even in the physical sense, another modern stereotype is the teenager holed up in their bedroom, awake until the early hours of the morning playing video games—one could perhaps draw the comparison between O'Connell's hungover crawl to join his son and wife in play, and the teenager's exhausted determination to stay awake just a little longer, to engage in play, exploration, socialising, or whatever virtual activity it may be. Although O'Connell makes a compelling point overall, towards the importance of not allowing technological advancement to diminish our human experience, I believe that we can be aware of the dangers of an ideal Transhumanist

---

[3] A famous example of this is a young disabled boy who despite being seemingly lonely and isolated, was discovered to have lived a rich and sociable life online, through the game of World of Warcraft. See Schaubert (2019); Musings of Life Blogspot (2013).

future whilst discovering new ways in which the human experience can again be prioritised, be that virtually or otherwise.

6.6   Conclusion

Such research into, and use of, virtual reality technology has a variety of practical applications. Whilst, as discussed above, we are inescapably heading towards a cyborg-esque future, in which we continue to advance technology, and thus further integrate with technology—either intentionally or due to a natural predisposition to do so—we have ability to create and design newer forms of technology which take advantage of such natural predisposition, whilst mindfully avoiding the inevitable dangers and constraints of such an existence. Although virtual reality technology is admittedly already a useful form of technology, I believe that it necessitates further research into new forms of virtual environments which *aren't* simply for entertainment purposes, virtual environments which offer similar advantages to technological integration as posed by Transhumanists, whilst maintaining a sense of "presence" and humanity that *doesn't* facilitate the same level of fear and concern towards technological integration.

Two interesting issues arise from the above conclusion, however. Firstly, further research into the *reality* of virtual reality needs to be done; in order to fully understand and develop virtual environments which benefit us the most, we first must understand the limitations of what can and *can't* be made virtual. Further, although there is a mass of research into virtual reality as medical treatment for a variety of physical and mental illnesses (see Freeman 2017; Barchester 2017; Gershon et al. 2004; Laver and Lange 2015), there is very little research into how consistent virtual engagement interferes with our capacity for real world engagement. Whilst much has been said on technology in general, often in the form of scathing articles towards social media and its demonic ability to turn younger generations into screen-enslaved zombies (see Engelhardt et al. 2011; Magid 2018; Taylor 2012), little has been said towards virtual reality technology which actively aims to immerse the user into a virtual environment, in which they are seemingly as free to act, think, and feel as their real-life human form. This increased level of immersion, and thus increased engagement with technology, seemingly—paradoxically—appears to cause *less* issues than its less immersive counterpart technologies, as argued above, and it is interesting to consider *why* this is the case, and whether or not this could continue to be the case in other circumstances. The second question which arises relates more directly to Clark's *Natural-Born Cyborgs*, and the issue of extended cognition as a whole; inasmuch as we can aim to understand virtual reality technologies and how best to design, develop, integrate and use them, we should *also* aim to understand the cognitive status as such, in order to complete the former tasks better. Whilst Clark, and many other philosophers of mind, cognitive scientists, and so on, continue research into the way that elements of the environment can be considered constitutive of mind, and even how our minds appear to have evolved in such a way that they are inevitably drawn to technological integration, further research needs to be made into the way that these mental connections with technology are impacted *in* a highly immersive situation. Even issues pertaining to embodied cognition become skewed as we face the potential of virtual bodies, embedded cognition becomes more complex as we face a *digital* environment within which we can live and act. This is particularly important as the crossover between human and technology, and technology and brain, becomes wider and more complex—and could ultimately offer a unique take on the Transhumanist ideal of the "evolved" human, or at least a greater understanding of what it may mean to direct our cyborg-predisposed brains towards more immersive technologies.

References

Barchester. 2017. Virtual reality offers respite for bedridden people. https://www.barchester.com/news/virtual-reality-offers-respite-bedridden-people. Accessed 21 Feb 2021

Bavelier, D. 2012. Your brains on action games. TEDx Talk, Lausanne, Switzerland. https://www.rochester.edu/news/video/bavelier-tedx-2012/. Accessed 16 Feb 2021

Carr, N. 2008. Is google making us stupid? The Atlantic. https://www.theatlantic.com/magazine/archive/2008/07/is-google-making-us-stupid/306868/. Accessed 16 Feb 2021

Carr, N. 2010a. The shallows: What the internet is doing to our brains. W.W. Norton, New York.

Carr, N. 2010b. The web shatters focus, rewires brains. Wired. https://www.wired.com/2010/05/ff-nicholas-carr/. Accessed 16 Feb 2021

Chalk, W., and Powell, S. 2020. Animal crossing: Why people play a game about 'meaningless tasks' BBC News. https://www.bbc.co.uk/news/newsbeat-52135528. Accessed 25 Feb 2021.

Clark, A. 2004. *Natural born cyborgs.* Oxford University Press.

Clark, A., and D. Chalmers. 1998. The extended mind. *Analysis* 58 (1): 7–19.

Clynes, M., and N. Kline. 1960. Cyborgs in space. *Astronautics*: 26–27

Engelhardt, C., B. Bartholow, G. Kerr, and B. Bushman. 2011. This is your brain on violent video games: Neural desensitization to violence predicts increased aggression following violent video game exposure. *Journal of Experimental Social Psychology* 47 (5): 1033–1036.

Freeman, D. 2017. Virtual reality in the assessment, understanding and treatment of mental health disorders. *Psychological Medicine* 47 (14): 2393–2400.

Gershon, J., E. Zimand, M. Pickering, B. Rothbaum, and L. Hodges. 2004. Virtual reality as a distraction for children with cancer. *Journal of the American Academy of Child and Adolescent Psychiatry* 43 (10): 1243–1249.

Greenfield, S. 2003. *Tomorrow's people: How 21st century technology is changing the way we think and feel.* London: Penguin Books.

Greenfield, D. 2007. The addictive properties of internet usage. In *Internet addiction: A handbook and guide to evaluation and treatment* (Chapter 8)

Heim, M. 1993. *The metaphysics of virtual reality.* USA: Oxford University Press.

Heim, M. 1998. *Virtual realism.* USA: Oxford University Press.

Kelly, K. (1994). *Out of control* (p. 176). Reading, Mass.: Perseus Books

Kennett, T. 2015. Why do we play video games that feel like work? Vice. https://www.vice.com/en/article/4x38aq/why-do-we-play-video-games-that-simulate-work. Accessed 25 Feb 2021.

Laver, K., and B. Lange. 2015. Virtual reality for stroke rehabilitation. *Cochrane Database of Systematic Reviews* 11 (1).

Magid, L. 2018. Majority of teens say they spend too much time on cellphones. Forbes. https://www.forbes.com/sites/larrymagid/2018/08/22/majority-of-teens-say-they-spend-too-much-time-on-cellphones-and-parents-agree/. Accessed 21 Feb 2021.

Musings of Life Blogspot. 2013. http://musingslif.blogspot.com. Accessed 20 Feb 2021.

O'Connell, M. 2017. *To be a machine*. London: Granta Books.

Pokemon Go. 2016. Mobile game by Niantic.

Richtel, M. 2003. The lure of data: Is it addictive? The New York Times. https://nyti.ms/2kFaPv1. Accessed 16 Feb 2021.

Richtel, M. 2010. Growing up digital, wired for distraction. NYTIMES.com. https://nyti.ms/2G3 SNIZ. Accessed 16 Feb 2021.

Salehan, M. 2013. Social networking on smartphones: When mobile phones become addictive. *Computers in Human Behaviour* 29 (6): 2632–2639.

Schaubert, V. 2019. My disabled son's amazing gaming life in the World of Warcraft. BBC News. https://www.bbc.co.uk/news/disability-47064773. Accessed 20 Feb 2021.

Taylor, J. 2012. How technology is changing the way children think and focus. Psychology Today. https://www.psychologytoday.com/us/blog/the-power-prime/201212/how-technology-is-changing-the-way-children-think-and-focus. Accessed 16 Feb 2021.

Vita-More, N. 2013. *The transhumanist reader: Classical and contemporary essays on the science, technology, and philosophy of the human future*. Chichester: Wiley-Blackwell.

Wallsten, S. 2013. What are we not doing when we're online. *Economic Analysis of the Digital Economy* 55–82.

Waytz, A., and K. Gray. 2018. Does online technology make us more or less sociable? *Perspectives on Psychological Science* 13 (4): 473–491.

Belinda Marshall is a Philosophy Ph.D. candidate at the University of St. Andrews as part of the SASP program with the University of Stirling, supervised by Prof. Michael Wheeler, following completion of an MRes degree in Philosophy from Nottingham Trent University. Her Masters' research focussed on phenomenological approaches to Virtual Reality technologies, from which her current thesis is focussed on an exploration of the links between virtual reality and theories of extended cognition and consciousness. Her main research areas are philosophy of technology (with a primary focus on virtual reality), phenomenology (including postphenomenology), and modern approaches to philosophy of mind (particularly 4E cognition). Most recently she has published her paper 'Being-in-the-Virtual-World' for the British Society of Phenomenology podcast, following several other conference presentations on topics such as 'Feminism and the Extended Mind' and 'Authenticity and Virtual Reality', for a range of institutes, such as the Leverhulme Centre for the Future of Intelligence, and the British Peronist Society.

Chapter 7
The Transhuman in the Workplace: Maximising Autonomy and Avoiding the Tyranny of Optimisation

Evie Kendal

Abstract The goals of the transhumanist movement are to promote wellbeing by transcending the limitations of human biology. However, within a neoliberal capitalist society, such a goal might be co-opted by corporate interests and lead to the exploitation of both enhanced and unenhanced workers. With its core principles including promoting autonomy for individuals and justice for all members of societies, the field of bioethics is well-suited to evaluate the promises and perils of transhumanist interventions in human existence, and their potential role in the workplace of the future.

7.1 Introduction

With its central focus on transcending the limitations of human biological existence, transhumanism is often equated with human enhancement (Bostrom 2005). Many transhumanist goals overlap those of biomedicine, such as reducing pain and suffering and extending life (More and Vita-More 2003). Where medicine and transhumanism diverge is a matter of some contention, particularly within the field of bioethics, where issues regarding the distinction between treatment and enhancement and the proper role of science and technology in human flourishing are frequently debated. This chapter will consider how two dominant principles of Western biomedical ethics—autonomy and justice—exist in constant tension when considering the issue of transhumanism. After addressing some common concerns about interfering with nature, and the potential that allowing human enhancements will promote inequality and exploitation, this chapter will continue with a defence of transhumanist interventions on the grounds of individual autonomy and liberty. Practical solutions to avoid exploitation will then be considered, from the perspective of both enhanced and unenhanced humans in the workplace. Ultimately, the chapter will demonstrate that any support of transhumanism on the grounds it maximises autonomy and personal

E. Kendal (✉)
Swinburne University of Technology, Melbourne, Australia
e-mail: ekendal@swin.edu.au

© The Author(s), under exclusive license to Springer Nature Switzerland AG 2022
E. Tumilty and M. Battle-Fisher (eds.), *Transhumanism: Entering an Era of Bodyhacking and Radical Human Modification*, The International Library of Bioethics 100,
https://doi.org/10.1007/978-3-031-14328-1_7

choice, carries with it the necessity of building in protections for equal opportunity and authentic human desire.

7.2 Transhumanism and Bioethics

In his influential article "A History of Transhumanist Thought," Nick Bostrom (2005)claims practical ethics experienced a "comeback" as a field of philosophical academic inquiry in large part as a response to medical ethics dilemmas brought on by scientific and technological developments in modern clinical practice. By the 1960s, medicine's self-regulatory mechanisms were already the subject of wide criticism, with the public's trust having been materially damaged by revelations of widespread unethical human experimentation over the previous decades (Wilson 2013). It was against this backdrop that medical and research ethics standards continued to strengthen in force. The formalisation of bioethics as a subfield of practical ethics was particularly triggered by advances in genetic and reproductive biotechnologies, such as in vitro fertilisation (IVF), combined with existing concerns about the beginning and end of life, allocating health resources, and dealing with patients lacking legal capacity to consent to medical care (Bostrom 2005). Bostrom continues:

> In the 1970s a broader kind of enquiry began to emerge... This field became known as bioethics. Many of the ethical issues most directly linked to transhumanism would now fall under this rubric, although other normative discourses are also involved, e.g. population ethics, meta-ethics, political philosophy, and bioethics' younger sisters – computer ethics, engineering ethics, environmental ethics (17).

The issues at stake include if and how various transhumanist goals, such as life extension, human biological, physical and cognitive enhancement, and biotechnological/prosthetic augmentation should be pursued. Applying a bioethical lens is particularly useful for examining these issues, as the arguments and principles of this field derive from practical consideration of the culture of medicine, and integration of science and technology into broader social structures.

A common practice in bioethics is to apply a principle-based approach to evaluate whether a proposed intervention should be prohibited, permitted and/or promoted. Adopting a principlist perspective (Beauchamp and Childress 2001) when considering transhumanist interventions, I argue the two most relevant ethical concerns are promoting and protecting autonomy and justice. To explore this claim further, I will now consider some objections to human enhancement that mostly stem from justice concerns, before providing a defence for transhumanism on the grounds of promoting autonomy. It will be my contention that support for transhumanism often involves a struggle between individual liberty and collective responsibility, with the other two main principles from Beauchamp and Childress's 1979 *Principles of Biomedical Ethics*, namely, beneficence and non-maleficence, balancing in the middle.

7.2.1 The So-Called "Wisdom of Repugnance"

I've placed the "yuck factor" argument against human enhancement first, as it is typically the easiest to overcome. This argument is articulated in Leon's Kass's work "The Wisdom of Repugnance," in which he claims, "repugnance is the emotional expression of deep wisdom, beyond reason's power to fully articulate it" (Kass 1997, 20). According to this argument, feelings of discomfort or revulsion when faced with the ideas or images of transhuman interventions may be a proxy measure for widespread and justified social disapprobation for these practices, even when there may be a lack of logical reasoning behind them. However, even Kass admits that "revulsion is not an argument" and that what society regarded as disgusting in the past may not be considered disgusting now (Kass 1997, 20). Examples include diverse sexualities and non-traditional family structures that would once have elicited the "yuck" response among the majority, but that are widely tolerated and even celebrated today. The same may one day be true for genetically engineered humans, cybernetic implants and the use of cryonics.

Most biotechnological developments have initially met opposition on the grounds of such "yuck" reactions, including organ donation, IVF, use of stem cells, and genetic research, to name a few (O'Carroll et al. 2011; Niemelä 2011). Had such intuitions against these practices been followed, many avenues for human flourishing, and indeed survival, would have been closed off. Rather than these feelings of disgust representing "which direction our ethics should follow," as Kass suggests, or standing in for some form of collective wisdom, Jussi Niemelä (2011) suggests "they have just as much to do with wisdom or ethical thought as a knee-jerk reflex" (274, 279). For Niemelä and others, the "yuck factor" as applied to emerging science and technology is a violation of its naturally evolved function: to protect humans from sources of biological hazard, such as open wounds or decaying corpses. He concludes:

> It appears that things that are not easily understood by utilizing folk-theoretical thought create a fertile soil for argumentation that strives to cause fear and disgust. Kass' arguments that strive to sanctify the natural prejudices biotechnology arouses are a good example of this rhetoric. … Even so, being aware of these biases may pave way for a better dialogue, in which unexplained fears and prejudice are no longer held as valid arguments used to craft legal and ethical guidelines (279).

There are many transhumanist interventions that may arouse the "natural prejudices" mentioned above, but separating out fear and misunderstanding from genuine ethical concerns requires critical engagement with the nature and purpose of these enhancements. Thus, the mere presence of widespread discomfort regarding transhumanism is not grounds to disregard the potential benefits this movement might yield for those choosing to engage with it.

A related objection to the "yuck factor" is focused on transhumanism being supposedly "unnatural" and therefore morally impermissible. This argument may arise most frequently in religious circles, but it has many sympathisers in secular society as well. Peters et al. (2008) label this kind of ethical reasoning a "*human protection framework*," noting terms like "*nature protection* or *anti-brave new world* or *anti-playing God*" might also apply (61). As will also be seen in the next section regarding the treatment versus enhancement debate, the core issue this kind of argument faces in bioethics relates to its reliance on a universal concept of "normal." If a transhumanist intervention can render a human body "unnatural," this infers both the existence of some uncontested classification of the "natural" human form, and its inherent desirability, or even moral superiority. This chapter will demonstrate how such views are practically unhelpful and potentially discriminatory.

At the heart of these bioconservative objections are perceived threats to "human nature." If, as Kenneth Mossman (2012) suggests, transhumanism is "a transition process from the present-day human to a more evolved posthuman," this may well threaten traditional views of "the essence of being human" (229). Naming social philosopher Francis Fukuyama as one taking a "decidedly bioconservative view," Mossman notes this perspective is often concerned with the potential attack to liberal democracy that unequal access to enhancing technologies might bring (229). As the basis of this democracy is human beings' fundamental equality under nature, altering this situation may undermine these core social values. However, I argue that equality does not denote sameness, and if natural biological variation has not led to the fall of liberal democracy, there is no reason to assume transhumanism will. As a society, we already accept many different levels of ability, and it is also likely that many transhumanist interventions might be "enhancing" in some circumstances, but neutral, or even disabling in others. For example, the near field communication (NFC) chips some biohackers have had implanted in their bodies for contactless payment and data storage, can under some conditions experience interference or be susceptible to cyberattack, and in the absence of another NFC-enabled device, these implanted chips merely become the functional equivalent of a subdermal piercing (Olenik et al. 2021; Mayers 2018). And just as some people find cosmetic body piercing distasteful, there will be those who find chip implanting unnatural or disgusting.

While the "yuck factor" and appeals to "naturalness" may not justify the prohibition of transhumanist interventions per se, they may serve a purpose in regulating some of the more extreme proposals, on the grounds of preventing physical and social harm. From the principlist perspective, such regulation may violate an individual's autonomy to select extreme body modifications but can appeal either to notions of justice, including social justice, or non-maleficence. The former argument will be pursued in more detail in subsequent sections, with reference to promoting equality and fairness in society, whereas the latter may represent an example of justified paternalism, e.g. where an individual's autonomy is overridden to protect them from causing harm to themselves. Such regulation is well tolerated in other areas of biomedical science and technology development, through imposing rigorous safety standards on medical devices and therapies. These restrictions remain even if someone voluntarily consents to a highly risky intervention.

7.2.2 The Treatment-Enhancement Distinction

The second objection to transhumanism commonly found in bioethics literature that I wish to address here is more complex than the previous one but similarly does not hold up under scrutiny. This is the argument that medicine's proper role in society is to return and/or preserve "normal" functioning, rather than seek to exceed human biological limitations This argument relies on a distinction between "treatment" and "enhancement" (Daniels 2008). Mossman (2012) notes this boundary is fundamental to discussions of human enhancement but that it remains "fuzzy" in practice, suggesting there is a "medical-intervention spectrum" that spans from treatment to prophylaxis to enhancement:

> Where on the spectrum a particular intervention falls depends on medical status (i.e., how we define normal, presence, or absence of disease) and the goals of the intervention. Therapy is generally defined as the management of disease. In the curative sense, therapy is the return to normal structure and function. Enhancement is defined as the alteration of characteristics, traits, and abilities beyond what is considered to be normal. The distinctions between therapy and enhancement rely on notions of disease, normalcy, nature, and naturalness. There is no bright line that separates therapy from enhancement. What is enhancement for one may be therapy for another. (230).

Some examples include pharmaceutical psychostimulants developed to manage attention deficit disorders, like ADHD, that may be used for cognitive enhancement by non-affected individuals (Mossman 2012). Studies suggest more than a third of students in American universities are using so-called "smart drugs" like Ritalin, Adderall, and Modafinil to enhance their capacity to study, with rates among those in the most competitive fields rising even higher to an estimated 43% (Le Dévédec 2020; Viet 2018). Of direct relevance to the workplace, such off-label use is also prevalent in high-pressure occupational environments, or where long periods of concentration need to be sustained, such as is the case for airline pilots, truck drivers, and surgeons (Le Dévédec 2020). Walter Veit (2018) notes that some even argue that when others' lives are at stake, such professionals may have an ethical obligation to use cognitive enhancements to combat fatigue (Viet 2018). It seems that for the patient with ADHD or narcolepsy, psychostimulants may be treating a biological pathology, whereas, for the doctor or truck driver facing a double shift, these same drugs might be alleviating the effects of a social pathology: a culture that promotes unsafe and exhausting working conditions. Mossman (2012) notes:

> Bioethical arguments raise serious questions about implementation of enhancement technologies but fall short of providing a clear, practical distinction between enhancement and therapy. Part of the difficulty is that therapy and enhancement share several common goals including fulfilling human desires and easing suffering (231).

The desire for wakefulness and attention, and relief from excessive sleepiness may be equally strong in the two groups above, but the methods of achieving these objectives might find themselves on opposite poles of the treatment and enhancement spectrum depending on which perspective is taken.

To demonstrate the false dichotomy between treatment and enhancement one need only look at some common examples from both categories and how they fit within transhumanism. According to McNamee and Edwards (2006), transhumanism is a "quasi-medical ideology" that promotes both therapeutic and human enhancement goals (513). By some definitions, transhumanists are merely a group of people pursuing human flourishing through targeted use of science and biotechnology, exercising their consumer power to achieve biological or cognitive modifications of their choice to improve their lives (McNamee and Edwards 2006). The overlap with the goals of modern medicine are immediately apparent. Some questions that arise, however, include: are medical interventions approved for therapeutic use also safe to use non-therapeutically, given the different risk/benefit profiles such uses entail? Should market demand be the deciding factor in whether a technology is made available for both curative and enhancing purposes? And will attempts to delineate a treatment/enhancement boundary lead to discrimination against those drawn inside and outside contemporary cultural constructs of health and disease?

To start unpacking these questions, we can first look at existing technologies that straddle this treatment and enhancement divide. The one several authors note as most salient to the workplace is the use of cognitive enhancers to maximise productivity through enhancing memory, concentration, and focus (Le Dévédec 2020; Viet 2018). Prescribed for diagnosable attention deficit disorders and conditions of excessive sleepiness, it is difficult to draw a firm line between pathological inattentiveness or fatigue, and normal human variation, particularly given the significant impact of lifestyle on both. As is often noted in the literature, other methods of enhancing cognition, such as caffeine, have been normalised in society and the workplace, while mood and concentration enhancing practices like meditation, mindfulness, and hatha yoga, are so well integrated into modern culture they are being used for "attention training" in primary schools (Reveley 2015, 805). James Reveley (2015) claims the latter is a form of "diffuse human enhancement strategy…whose purpose is to pathology-proof the younger generation," including through "preventing the onset of depressive symptoms," for which such interventions have proved efficacious (805, 808). Thus, he claims mindfulness programs in schools are proactive enhancement interventions focused on promoting wellbeing, creativity, and attention, rather than "reactive therapeutic responses to dysfunction," as would be the case for psychostimulants prescribed to achieve the same outcomes (814).

Some may argue there is a qualitative difference between meditating or mainlining coffee to finish a school or work project, and off-label use of pharmaceutical psychostimulants, based on the risk of side-effects. However, studies suggest caffeine intake influences blood glucose levels and insulin sensitivity and can cause significant sleep disturbances in those with caffeine sensitivity (Dewar and Heuberger 2017; Clark and Landolt 2017). Similarly, so-called "trauma-informed" or "trauma-sensitive" yoga practices were developed in response to anecdotal evidence that traditional versions were emotionally overwhelming for participants with a past history of trauma. Returning to Mossman's (2012) "medical-intervention spectrum," there does not seem to be a clear-cut method of distinguishing between therapy, prophylaxis, and enhancement with regards to the use of psychostimulants, mindfulness, and other forms of attention improving interventions.

The next potential targets for transhumanists are prosthetics and implants. Here distinctions between treatment and enhancement are similarly complicated. Most would agree an artificial limb provided to someone whose own did not develop fully or was severely injured, falls within the scope of "treatment"—it is, after all, returning someone to a typical level of human functioning based on an assumption regarding the number and type of limbs such bodies usually have. However, the materials prosthetics are made of may be stronger than their organic counterparts, e.g. human bone. Does this then render the user an enhanced (trans)human? Or does circumscribing the idea of a "typical" human body in the first place on diverse and differently-abled bodies represent discrimination, or what David-Jack Fletcher (2014) labels the "mass disenfranchisement of clusters of non-hegemonic bodies" (90). With particular reference to transhumanist genetic engineering and the potential "reinscription of eugenic regimes" such practices could engender, Fletcher (2014) claims such attitudes invalidate "different modes of existence for the goal of perfecting the human race" (90). If a prosthetic limb is distinguished from a biotechnological augmentation solely by casting the differently-abled body as diseased and in need of therapeutic intervention, this objection seems to hold. This will be particularly the case when the prosthetics used for "treatment" are identical to those transhumanists repurpose for "enhancement."

The same issues arise when trying to differentiate between widely used medical implants, such as pacemakers and cochlear implants, and transhuman "insideables"—so-called "in-the-body technologies" implanted for non-therapeutic purposes (Fox 2018). The former, while common, are not uncontroversial, particularly the cochlear implant. This device is surgically implanted into the middle ear of the recipient to allow the perception of sound through stimulating nerve endings in the inner ear. Whether this constitutes a restoration of "normal" functioning, or a concentrated attack on Deaf culture as a cultural and linguistic minority, depends on who you talk to (Sparrow 2010). Again, we are faced with a technology some consider curative, others view as a (relative) enhancement, and some consider disabling, depending on context. The usual analogies with "enhancing" devices like spectacles for people with vision impairments also apply here.

As enhancement technologies typically stem from therapeutic advances, and it is difficult to establish a firm line between them, I argue we cannot attempt to use the treatment and enhancement distinction to block transhumanist endeavours. That risk/benefit analyses will be different for different uses of technology is also not a compelling reason to ban non-therapeutic applications, as this is already the case between different patients accessing the same technologies for treatment and can be managed in a similar way. The beneficent goals of medical research and development should not be impeded because the achievements they bring cannot be disentangled from non-therapeutic uses. For example, it would be morally reprehensible to stall development in the field of prosthetics for people suffering catastrophic injuries, just because the devices created might have a secondary function within transhumanism of enhancing "normal" bodies' strength or range of function. But this does not necessarily mean we must embrace a free-for-all attitude toward human enhancement, as there are legitimate concerns about equal access and potential social harms, as will be discussed now.

7.2.3 Fostering Inequality: The H+ Haves and the H Have-Nots

The potential for transhumanism to split the human race into the enhanced (H+) and unenhanced (H) is perhaps one of the more compelling objections found in bioethics scholarship. What is fascinating about this particular argument against human enhancement is the way it highlights types of inequality that we routinely tolerate, by contrasting them with those we typically do not. While it is understood that the offspring of wealthy parents have access to many advantages of education and opportunity that other children do not, these are rarely considered unfair "cognitive enhancements." However, when identical advantages might be conferred through high-cost genetic engineering, this is often considered unjust. Jeff Pruchnic (2016) claims this concern also overlaps with the previous issue regarding "unnaturalness," with genetic engineering supposedly representing "an unethical encroachment of technoscientific practice on the 'natural' process of human reproduction," in our attempts to create "better humans" (19).

According to Tod Chambers (1999), bioethics scholarship, like all genres, has its own folkloric shorthand. He notes that within this field, the name "Dax," for example, a reference to severe burns survivor, Dax Shepard, can stand in for an entire ethical argument about respecting the autonomous wishes of patients, even to refuse life-saving interventions (4). When it comes to discussing the possibility that transhumanist technologies, such as genetic engineering and cognitive enhancement, will foster inequality between enhanced and unenhanced humans, I argue the shorthand "haves and have nots" is the one that often comes into play. Outside of the field of bioethics, the phrase may be invoked to denote any two groups in which a vast disparity in wealth exists, but interestingly, it is not typically used in bioethics to discuss differences in material benefits per se, but rather differences in *access to biomedical benefits*, including many that might be considered transhumanist. That we may discuss the "genetic haves and have nots" when decrying the unfairness of allowing the wealthy to select characteristics for their so-called "designer babies," while we do not use the same language to discuss the "Lamborghini haves and have nots," or the "Rolex haves and have nots" serves to illustrate how health and medical interventions are treated differently to other resources, and indeed wealth itself. That members of society lack equal access to Lamborghinis and Rolexes on the basis of socio-economic status is tolerated, while the idea of some aspects of healthcare and biological existence becoming luxury commodities remains ethically problematic. While Pruchnic (2016) claims some such interventions previously considered "unassimilable to processes of commodification" have now become market products "to be 'consumed' in contemporary economic exchange," the prevalence of distributive justice concerns regarding genetic engineering of offspring or life extension processes, for example, indicate this assimilation is by no means complete or uncritically accepted (29). This is an important consideration when determining appropriate limits to libertarian demands for enhancing technologies driven by market values.

At the core of the "haves and have nots" debate is the principle of justice. In his article, "Cognitive Enhancement and the Threat of Inequality," Veit (2018) claims there is "no empirical evidence that cognitive enhancement will lead to more inequality," claiming it instead "has the potential to decrease it" (405). Similarly, he notes genetic enhancement could "get rid of a major source of inequality, i.e., the elimination of the natural lottery" (407). Even if evidence of increased social inequality did exist, Veit claims it is by no means certain this cost would outweigh the potential benefits enhancement might confer:

> there are other areas where human enhancement can potentially increase well-being enormously and outweigh losses through inequality. Even if only one cognitively enhanced scientist discovers a new cure to a disease, the effects on well-being would be substantial. Even minor increases in productivity, say 10%, in the workforce could have substantive positive effects in the long-run (408).

Veit claims concerns about inequality regarding human enhancements are more accurately issues of unequal access. This is precisely the issue that fuels the "haves and have nots" objection in bioethics, an example of which is seen below from Stephen Fox (2018):

> Techno-utopia propositions include arguments that technological advances should bring about utopias where human beings transcend the constraints of biology and become transhuman or even posthuman. However, this techno-utopia is regarded by some as a very dangerous idea that will actually lead to a techno-dystopia of enhanced "haves" and unenhanced "have-nots" living in societies that are more polarized than before widespread human enhancement (54-55).

This threat to equality and social harmony challenges many transhumanist goals. Stephen Lilley (2012) claims egalitarianism lies at the heart of much transhumanist thinking, with enhancements often considered a "social good" the state might have a duty to promote or at the very least attempt to prevent the conflict that "might arise if only the affluent have access to them" (11). Transhumanists like Nick Bostrom and Anders Sandberg have suggested attempts to prohibit enhancement technologies in order to prevent inequality are likely to exacerbate the issue by driving transhumanist practices underground, where they will be even more expensive and inaccessible (Viet 2018, 409). Instead of becoming safer and cheaper over time, a black market would ensure risks and prices remained high, keeping enhancements out of reach for many. But what such proponents are mostly concerned with is promoting individual freedom to pursue transhumanist interventions, if desired. Which brings us to the next bioethical principle of interest here: autonomy.

7.2.4 Autonomy, Liberty, and Informed Consent

There is a joke often told to undergraduate bioethics students first learning about principlism: the four principles of bioethics are autonomy, autonomy, autonomy, and autonomy! While hardly a knee-slapper, the joke does communicate one key fact: modern biomedical ethics is deeply concerned with the issue of personal liberty and self-governance. How the principle of autonomy is usually enshrined in research and clinical care is through codes and standards that promote freedom of choice and informed consent, e.g. for medical treatment or research participation. Advocates for allowing the expansion of transhumanist interventions often appeal to concepts of the good life being dependent on personal freedom and choice, as McNamee and Edwards (2006) note:

> For one group of transhumanists, the good is the expansion of personal choice. Given that autonomy is so widely valued, why not remove the barriers to enhanced autonomy by various technological interventions? ... One extension of this line of transhumanism thinking is to align the valorisation of autonomy with economic rationality, for we may as well be motivated by economic concerns as by moral ones where the market is concerned (515).

These authors claim such supporters see the transhumanist project merely as "a way of improving their own life by their own standards of what counts as an improvement" (514). This idea that individuals are best placed to know what medical interventions align with their personal goals and values is at the core of patient-centred care, a treatment philosophy that arose in response to both the medical paternalism of the past in terms of clinical practice, and the human rights violations seen in unethical human experimentation (Kwame and Petrucka 2021). For transhumanism, autonomy is enhanced through greater availability of biotechnological interventions for personal selection and purchase.

We have already seen that even if the majority find transhumanism distasteful or simply unappealing for themselves, that this does not justify prohibiting others from accessing interventions they may personally benefit from. This is perhaps particularly the case where curative and enhancing technologies cannot be clearly distinguished. As noted in the previous section, concerns about human enhancement fostering inequality in society are valid, but if we apply John Stuart Mill's (1859) "harm principle," I argue we can satisfactorily meet both demands. According to Mill, in his famous text, *On Liberty*:

> the only purpose for which power can be rightfully exercised over any member of a civilised community, against his will, is to prevent harm to others. His own good, either physical or moral, is not a sufficient warrant (223). [sic]

With reference to transhumanism, this argument asserts citizens should be free to choose what they do with their own bodies up until the point these choices impact on the rights of others. As will be seen later in this chapter, one such avenue for potential impact is in the sphere of employment inequality. However, using Mill's formulation, the threshold for establishing when someone's personal choice for engaging with transhumanist interventions has infringed on the liberty of others in society, must be relatively high. I will argue that most of the potential future issues arising in employment inequality between enhanced and unenhanced humans are better managed through equal opportunity legislation, that would also strengthen existing protections for employees with disabilities.

The principle of autonomy is not just about freedom of choice, but also appreciation and voluntary acceptance of the consequences of said choices. This is where informed consent enters the picture. For Marion Grau (2003), some of the current techno-fetishism seen in post 9/11 America can be attributed to "silver-bullet-ism" and overreliance on technology "as the salvific solution" to widespread feelings of apocalyptic terror (144). She argues that rather than seeking solace in religion, there has been a shift toward placing faith in "eclipsing" the body and surpassing the limitations of humanity as the way to restore public safety (144). However, if the pursuit of transhumanism is motivated by unrealistic beliefs about the role of technology in society, this undermines autonomous decision-making. Thus, if we are defending the provision of enhancing technologies on the basis they promote liberty and autonomy, this requires that those pursuing them are provided with the necessary knowledge to understand what they are choosing and its expected scope of impact on their lives. This is another reason to prevent widespread prohibition of human enhancement, as regulation of consent procedures cannot be achieved if the majority of enhancements are only available on the black market. This brings us to the next consideration: that whether or not they are legal, many believe that human enhancements are inevitable.

7.3   The Inevitability of Enhancement

While it is a powerful influence, promoting autonomy is not a universal defence for pursuing transhumanism. Jean Bethke Elshtain (2004) notes dominant Western cultural ideals of "choice, consent, control" are often invoked when discussing interventions intended to overcome human biological limitations, promoting an assumption that technological development is inevitable and "absolute self-possession" through advances in biotechnology is both possible and desirable (165). But is transhumanism really inevitable? Given its relationship with modern medicine and neoliberal economics, I argue that at least to some extent it is.

Marchant and López (2012) believe that because most enhancement technologies develop as extensions of treatments for supposedly "legitimate health problems," that it will be "politically difficult, and perhaps ethically unacceptable" to try and stop their development (262). These authors also note that any strategy intended to monitor compliance with human enhancement restrictions are likely to "unduly intrude on and burden individual freedom," and if imposed at a national level would simply drive medical tourism, where those with the resources obtain treatment overseas (262). The commercial potential for a significant global market in enhancement technologies is already understood by many advocates of transhumanism, and Mossman (2012) notes market forces dictate that if transhumanist interventions are only prohibited

in some jurisdictions, "other nations will aggressively pursue research and development," to capitalise on this potential economic gain (229). Coming at the issue from
the opposite end, Veit (2018) notes attempts to prohibit transhumanist interventions
will face logistical challenges, for how could a society punish those who violate these
restrictions? Would someone who was "illegally" enhanced be forcefully reverted
to their pre-enhanced state? In the case of radical life extension technologies would
this reversion be tantamount to state execution? Financial disincentives are unlikely
to have sufficient force to dissuade those intent on accessing transhumanist interventions, for as the saying goes: punishable by fine means legal for the rich. Drawing
an analogy between banning human enhancement and the prohibition of alcohol,
Veit (2018) notes, "a world where it is illegal does not equate a world where this
technology is not available," but rather a world of increased illegal activity (409).

Another reason transhumanism can be considered inevitable is enhancements are
becoming increasingly achievable through DIY biohacking. Fox (2018) notes the
rise of "implant parties" where individuals gather to have "insideables" implanted
in their bodies, claiming, "[a]t these parties, they can share in social interactions
similar to those of Botox parties that preceded them" (51). I argue that what we
are seeing in both these "parties" is evidence of a new form of biocitizenship, where
people united by common interests in body modification and anti-aging are gathering
together to achieve these goals. Plows and Boddington (2006) note "biocitizenship"
is often a term applied to "disease communities" when using their collective force
to pressure governments or other organisations to listen to their concerns regarding
"access, price, quality and availability of treatments and cures—in other words, *over
increasing access to the fruits of biotechnology*" (121). This definition of biocitizenship as collective mobilisation over issues of health and access to technology could
also apply to transhuman communities. While they lack a disease or genetic risk
factor in common, which is often a feature of other biocitizen groups, the agitation
for regulatory change and greater access to biomedical interventions to promote the
needs of their interest group are comparable. As Katz and Gish (2015) note:

> Central to this biosocial vision of modifiable life, including aging, is that it can be improved,
> enhanced, and optimized beyond what is needed to sustain good health. And this is where
> ethical debates about the limits of enhancement and longevity find themselves today… the
> biosocial order is one that encourages people to congregate as biocitizens around issues of
> medical treatments and environmental risks … in our case, anti-aging medicine and cosmetic
> bodywork have also become powerful agents for animating biocitizens to pursue a positive
> aging science of restoration and repair in contrast to the negativity of geriatric medicine
> (42-43).

As many transhumanist interventions are now possible without engaging with
the medical sector, attempts to limit human enhancement through restricting, for
example, the kinds of biomedical devices a doctor can implant, are unlikely to curb the
"congregation" of transhuman "biocitizens" to access these technologies, including
at so-called "implant parties."

Returning to the issue of treatment versus enhancement, it is also important to
consider that many bioethicists do not oppose non-therapeutic use of medical interventions per se, but rather question the state's role in subsidising the cost of such

interventions for citizens (Daniels 2008). For example, few oppose the legal and safe provision of Botox for purely aesthetic reasons, assuming voluntary, informed consent is given, but many would argue users should have to cover the costs of such cosmetic interventions themselves. Leaving aside the issue of establishing when non-medical interventions may be life-improving, life-affirming, life-sustaining, or life-saving, including those that affirm gender identity or improve mental health, when applied to transhumanist interventions this distinction opens up various avenues for inequality. If the only limits for transhumanism become those of imagination and cost, this further validates the H+ versus H justice concerns discussed earlier. On the issue of biohacking and imagination, Fox (2018) notes:

Mass imagineering includes body hackers who imagine enhanced capabilities for themselves and put automated engineering devices inside themselves to realize what they imagine. Thus, they engage in a particular type of creative do-it-yourself (DIY). Enhanced capabilities include having automated auditory representations of colors, automated reading of the Earth's seismic movements, and automated notifications every time one faces north in order to become a human compass (52).

This quote demonstrates a number of the ethical concerns covered so far in this chapter. Imagining a situation in which the last intervention has been successfully achieved in a transhuman citizen, the following questions might arise: does becoming a human compass violate the essence of being human? Did the recipient have enough information about the consequences of this intervention on their life to give informed consent? Was the procedure illegal, and if so, on what grounds and what can we do about it now? Was the procedure conducted in a regulated and safe environment or a backyard operation? Who paid for it and was it a legitimate use of medical resources?

While many answers to the above questions will depend on the policy conditions in the relevant location, what I would like to move on to now is considering a new question: how might being a human compass, or having a host of other transhumanist interventions, impact this individual's social and employment opportunities? This also carries with it the central question of this chapter: how can we balance the principles of autonomy and justice when it comes to transhumanism in the workplace?

7.4 Promoting Human Flourishing for the Transhuman Employee

Many of the arguments in favour of allowing transhumanist interventions on the basis of promoting autonomy in general, also apply when considering the workplace, specifically. If it is plausible that an individual might want to enhance themselves to achieve various life goals, it follows they may also seek enhancement to achieve certain career and occupational goals. According to Peter Bloom (2020), transhumanist enhancements can "serve the needs of humanity even while unrecognizably transforming them" (12, 13). This section will focus on the potential positive transformations enhancement technologies might yield for future workers and the workplace, before the next section considers some of the pitfalls of embracing transhumanism within a neoliberal capitalist context.

7.4.1 Work Smarter, not Harder

For many transhumanists, enhancing drugs and technology represent a way to promote wellbeing and improve the human condition. Techno-utopian and transhumanist, Ray Kurzweil (2010), claims advances in molecular biology and super-intelligent computer systems will one day solve humanity's largest problems, eliminating poverty, environmental damage, pain, suffering, and mortality. He also notes science and technology industries will continue to create many jobs and generate significant wealth, that he believes will improve standards of living across the globe (652). When discussing the impact of technological development on employment, Kurzweil (2010) claims these benefits are set to increase exponentially once we abandon the limits of biological neuronal systems, claiming human brains "suffer from severe limitations" and our "version 1.0 biological bodies are likewise frail and subject to a myriad of failure modes" (41). While he seems to think the transcendence of human biological limitations is likely to be a species collective event, the hypothetical discussed earlier about the one cognitively enhanced scientist discovering a cure for a disease, reminds us that even if only a few people engage with enhancement, this has the potential to benefit many others (Viet 2018). Relatedly, Fox (2018) claims that the "hacker ethic" is often focused on making the world better, not just benefiting the self (52). When applied to biohacking, he suggests a goal like implanting night vision devices into everyone, could reduce the energy costs of providing public lighting, thereby saving power and reducing environmental damage (52). While it is highly unlikely everyone would want night vision technology implanted in their bodies, I argue this kind of intervention might be particularly valuable for people whose work requires high visual acuity at night, such as search and rescue crews. Just as some people opt for laser eye surgery rather than the inconvenience of wearing spectacles, it is plausible some employees might prefer enhancement technologies over the use of other occupational equipment, such as night-vision goggles.

When considering whether transhumanist interventions will be widely rejected if it can be demonstrated enhanced employees are likely to unfairly disadvantage their colleagues, Veit (2018) writes:

> I deem it unlikely that someone would be opposed to surgeons taking cognitive enhancers to increase their ability to concentrate during a complicated procedure, just because they would gain a competitive advantage over their peers. The value of saving lives outweighs the negative effects of peer pressure to take enhancers (408).

While intended to capture the idea of a communal benefit for an otherwise individualistic intervention, this quote nevertheless also highlights that there likely would be "competitive advantage" for the transhuman in this situation. Due to enhancement, they will be able to perform at a higher standard and for longer periods of time than their peers. From the perspective of the individual pursuing such advantage, transhumanist interventions might represent just one of a vast range of career "enhancing" opportunities, including educational interventions such as re-credentialing, extension studies, or additional training, or as mentioned earlier, activities like meditation and mindfulness, or use of substances like coffee or pharmaceutical psychostimulants. As with these other options, the individual is likely best placed to determine what interventions align with their personal view of human flourishing, including in their career.

Beyond competitive advantage against their human peers, Fox (2018) notes transhumanist interventions, such as implants and other cyborgian modifications, might give employees the ability to compete with the robots many fear are set to take over their jobs in the future. He further notes that through trial and modification to meet end-user needs, such enhancements would undergo "technology domestication," being adapted to individuals' preferences (55). In the workplace, such preferences might include being able to complete tasks faster and thereby increase the amount of leisure time an employee has in their day. Enhancements might also be viewed as a way of opening up employment opportunities in a similar way to education, with prospective employees weighing up the costs and benefits of transhumanist interventions in much the same way as they may the pecuniary and opportunity costs of completing a university degree. Individuals may engage with enhancement technologies that give them physical or mental capacities that are in high demand in the workplace as a way to secure employment or advance their careers. We can even imagine new jobs being created that capitalise on skills that have not previously been seen. Or, as will be discussed next, an employee may choose enhancement to minimise their risk of workplace injury.

7.4.2   Work Safer

Previously we've discussed the potential that for some professionals engaging with human enhancements might be morally praiseworthy, or even obligatory, in the interests of saving lives. Now we will briefly consider the potential safety benefits of enhancement for workers themselves. As a society, we have come to accept that some jobs are hazardous, and often compensate those people who choose to provide relevant services with hazard pay in recognition of the risks they expose themselves to in the course of their employment. This tolerance of risk has its limits, with employers held to standards of occupational health and safety that dictate all reasonable measures to prevent injury should be taken, including providing safety equipment and training. Notwithstanding these precautions, at present, those seeking the financial benefits of working in a hazardous industry have no choice but to submit their bodies to additional physical risks compared to workers outside of these industries.

Considering enhancements targeting occupational exposures, possible examples discussed in the literature include modifying skin to make it more radiation-resistant, altering metabolism to improve energy levels and reduce sleep needs, improving immune function, and optimising organ systems, e.g. for more efficient respiration, circulation, muscular contraction, etc. (Szocik et al. 2019). While some of these are intended to maximise performance, others are directly related to improving safety and preventing injury or illness. We can also imagine interventions that would allow workers to withstand extremes of temperature, prolonged time underwater or at high altitudes, and which could protect their skin and lungs from damage caused by environmental toxins. Enhanced physical strength or use of prosthetics could ease the burden of manual labour, while other interventions may make the body less susceptible to electrocution, burns, or infection. Our earlier example of becoming a human compass might aid in transportation and logistics jobs, while neuro-cognitive enhancement could improve concentration and mood stability for those engaged in high-stress or emotionally challenging careers. It is important to remember that many occupations carry routine risks for morbidity and mortality and that many of these have been normalised or are merely addressed through hazard pay. Transhumanist interventions may provide employees more options to protect their health and wellbeing without compromising their career goals or livelihood.

As the focus of this chapter has been transhumanism in the workplace, potential interventions on offspring have mostly been excluded. However, interventions into reproduction are directly relevant to many people's employment, with egg freezing and menopause reversing procedures becoming increasingly common for women wanting to avoid career interruptions at critical points in their working lives (Le Dévédec 2020). Cosmetic interventions to avoid age-based discrimination in the workplace have also been reported (Le Dévédec 2020). But these examples also highlight existing and future concerns regarding the motivations behind seeking enhancement, and how transhumanism might exacerbate forms of discrimination that are already present in employment today.

7.5 Protecting Worker Rights for the Enhanced and the Unenhanced

The justice concerns noted earlier in this chapter are perhaps most tangibly manifested in the work environment. Pruchnic (2016) notes concerns about unenhanced humans being "outpaced by their genetically enhanced counterparts" and the potential for "gene races" at the level of nations and corporations, represent some of the major ethical dilemmas brought about by human enhancement (19). Likewise, Mossman (2012) is concerned that attempts to correct human imperfections will lead society down the "slippery slope of medicalizing normal variations," noting normality and disease are "moving targets...bound up in personal and community views" (232, 234). Such distinctions have formed the basis of social and occupational discrimination, by implying that it is the individual's responsibility to "fit in" with the corporate environment, including through personally managing any perceived impediments to performance. According to Reveley (2015), this includes a supposed obligation to use neuroscientific therapies to improve brain function. For neuro-divergent individuals, such expectations may be unfairly burdensome or exclusionary, e.g. requiring someone with autism spectrum disorder to adapt to working in a loud or brightly-lit office where they experience sensory overload, or someone with ADHD to work in an open-plan office where they are liable to be frequently distracted. While such conditions may prove challenging for neuro-typical employees, they may be overwhelming for candidates with special needs. Any suggestion cognitive enhancement, rather than better workplace neuro-ergonomics, is the correct way to manage these issues is ethically problematic, and again relies on definitions of normalcy that are potentially discriminatory. Pressuring employees with real or perceived disabilities to engage with enhancement technologies to keep up with their peers, directly violates various principles of justice. Likewise, any coercion for employees with standard physical or mental capacities to accept transhumanist interventions, as will be discussed below. However, it is likely that there will be employees in the future, including some with disabilities, who will find the additional options transhumanism provides to be empowering, and a means of widening their scope of choices.

7.5.1 Protecting Equality of Opportunity

A central tenet of equal opportunity is ensuring that providing options for some people does not diminish choices for others. In terms of enhancement technologies, this means making sure varied bodies and levels of ability are supported in society, even when interventions are available to eliminate physical or cognitive differences. The existence of artificial limbs and cochlear implants does not necessitate their use among their target audience, and neither should the development of transhumanist interventions carry any obligation to engage with them. However, as Susan Smith (2016) notes, many disability scholars are recognising the drive for "posthuman self-expression" among people with disabilities, claiming their engagement with prosthetics and other technologies have always connected them with the cyborg (261). She sees the interdependence between the physical body and prosthetics to be an opportunity to consider different ideas of what it means to be human in the present time. However, there will also be many among disability communities and the general public who do not wish to engage with existing or future interventions aimed at enhancing physical or cognitive abilities, and these groups will need protection in employment legislation to avoid discrimination. I argue now is the time to legally enshrine such protections, before human enhancement in the workplace becomes common. The status quo currently favours H, rather than H+, and the continued use of affirmative action to promote the interests of minority groups in education and employment. Any failure of existing anti-discrimination laws for equal employment opportunities are evidence only that such measures need to be more strictly applied, and not that discrimination against the unenhanced in the future is inevitable.

Veit (2018) claims that since employers already fund "enhancements" for staff, such as advanced training or free coffee dispensers, if transhumanist interventions were shown to improve productivity they may be willing to fund those as well. At that point it would be necessary to ensure all employees were given fair and equitable access to subsidised enhancements, should they wish to avail themselves of them. He also envisages a future where people can take out enhancement loans in much the same way as they currently take out student loans, with the hopes of improving employment prospects in the long term (Viet 2018, 407). These methods might go some way toward equalising access across different socio-economic levels, with state subsidisation representing another possible economic model. What is most important in terms of protecting equal opportunity is ensuring wealthier citizens are not able to compound their advantages of birth or situation with advantages of enhancements that are unavailable to other members of their society. This is to ensure the unenhanced do not lose competitiveness in the workforce or feel pressured to pursue enhancement themselves just to keep up. Since promoting autonomy is the dominant defence for allowing transhumanist interventions to be developed and released, allowing such pressures to diminish autonomous decision-making to choose or refuse enhancement is antithetical to the purpose.

In terms of competition, as noted in the previous section, transhumanist interventions may stimulate the creation of new jobs heretofore not considered due to natural human limitations. If this were the case, competition for existing jobs may actually decrease for the unenhanced, as their enhanced counterparts move to other industries. Again, this highlights the need to protect current working conditions to ensure these jobs do not become devalued. Extending existing provisions for selecting candidates on the basis of their ability to achieve intrinsic features of the job would prevent employers from being held liable for discrimination for selecting enhanced employees for work an unenhanced candidate would be physically or cognitively incapable of doing.

I argue that many of the other concerns about transhumanism potentially leading to employment inequality are more of an opportunity to reflect on how we want employment to look in the future, than a genuine threat to equality. If we are concerned that allowing some people to reduce their need for sleep will effectively cut others out of employment, we are suggesting that we want to uncritically allow workplaces to extend work hours beyond that which is currently possible for an unenhanced human. As is seen in some occupations, this has already occurred, leading to existing safety concerns, e.g. for long-haul transportation workers. This demonstrates the need to re-evaluate existing workplace relations to prioritise worker wellbeing over profit generation. Which brings us to our next consideration: how can we avoid a future where transhumanism is co-opted for corporate gain?

7.5.2   Avoiding the Tyranny of Optimisation

Why hire a lowly human when you can hire a transhuman who doesn't sleep? In the absence of restrictions on shift length, such reasoning may lead to a worsening of work conditions for the enhanced and unenhanced alike. However, if we strengthen protections against workplaces imposing an unreasonable spread of hours on employees, this might prevent unsafe work practices while also promoting the freedom for employees to choose enhancement technologies because they genuinely desire them. In terms of avoiding disadvantage to peers, the ability to work 30 hours straight without rest, for example, will be of no benefit in a context where shift length is limited to 8 hours. However, it might reduce discomfort for those roles that already necessitate those kinds of hours, which may entice more people within these fields to consider human enhancement. But who decides what is "necessary" in terms of performance in a neoliberal capitalist context?

To explore the above question, this section will draw on Nicolas Le Dévédec's (2020) arguments from "The Biopolitical Embodiment of Work in the Era of Human Enhancement." For Le Dévédec, despite receiving minimal scholarly attention, the workplace is "at the front line of…technoscientific and biomedical transformations of the body" aimed at optimising employee performance (57). He questions:

> Is the augmented body an expression of a liberated body and of greater worker empowerment? Or rather, does it not signify the emergence of a body and a worker that is first and foremost adapted to the growing intensification of work conditions and of the pressures to perform and be productive that characterize advanced capitalist societies? (57).

Claiming enhancement as a new expression of biopower within capitalist society, Le Dévédec (2020) suggests this paradigm has the potential to create "new forms of exploitation and alienation" for workers, as it encourages the "internalization of the neoliberal norms of self-surpassment and performance" (57, 58). In this way he argues transhumanism may promote employee adaptability and compliance with increasingly intense work practices, rather than the "questioning of capitalist organization models of work" (58). Drawing analogies with nineteenth century Taylorist disciplinary models of management and the metaphor of the "human engine" whose output needed to be maximised through external pressure and punishment, he notes bioengineering could take control over workers' bodies even further, by conditioning them to internally push against their own physical limitations (64–67). He continues:

> The extension of biopower to the life itself of salaried workers – what Fleming calls biocracy – does not just affect their social life. It is also a new way of seeing employees' bodies and vital capacities, which are understood and utilized differently than in the classic biopolitical model. … Even if the bodies were transformed, conforming to the capitalist machine, as Marx (1867) clearly showed, this transformation was an effect and a result of working conditions rather than a goal in of itself. The human body could be incited, pushed and made to comply with the requirements and productivity norms, and thereby profoundly altered, but in no way could it be remodelled, remade and redesigned in itself via bioengineering (66-67).

However, with human enhancement technology, Le Dévédec notes "the body is no longer understood as an immutable given, like in Taylorism. It becomes wholly plastic, malleable and perfectible" (67). In other words, management no longer have to rely on social pressure to achieve conformity, they can co-opt their employees' biological systems, intellectual capacities and creativity to maximise productivity. Mossman (2012) notes that one of the largest sources of funding for enhancement research comes from the U.S. military, driven by a desire to improve "combat effectiveness" and "peak performance" of soldiers (233). That transhumanist interventions could be used to push workers or soldiers even harder, or just come to represent part of an employee's coping strategy for workplace pressure, reminds us that while technologies themselves may be neutral, their application in society rarely is.

If we accept that at least some transhumanist interventions are inevitable, as discussed previously, and that there is the potential that enhancing technologies may serve to disadvantage unenhanced individuals or exploit enhanced workers, it becomes clear that stronger employment protections need to be in place before transhumanism can be widely embraced in the workplace. This might include regulating what enhancements are available and on what grounds they can be accessed. More immediately, it might restrict what kinds of enhancements employers can subsidise or seek to promote among employees. This sort of regulation may impede some liberties, but Lilley (2012) claims this would not necessarily run afoul of transhumanist philosophy, as although transhumanists "favor entrepreneurial capitalism" they are not wholly supportive of economic libertarianism and may welcome some state influence on distribution of enhancing technologies among citizens (11).

Ian Barns (2018) notes that technology has been both essential to the development of neoliberalism, with its consumerism and environmental destruction, and our dreams of transcending this material existence and pursuing a "posthuman or transhuman future" (146). Transhumanist interventions have the potential to improve work-life balance for employees, through greater efficiency, and promote safety, as previously discussed. However, the attitude that human limitations can and should be overcome through the application of biomedical and scientific interventions, has the capacity to dehumanise workers and lead to a tyranny of optimisation where humans are valued predominantly based on their productive output. Reveley (2015) claims this is already being seen in schools where low-tech cognitive enhancements and attention training are promoting a new form of "human neurocapital," where children's attention is now considered a "prime, but friable, human resource" to be cultivated for future productivity (807, 814). Similarly, Le Dévédec (2020) claims transhumanism may promote more extreme exploitation of workers by promoting workaholism and providing a means through which natural limitations can be surpassed. In this way, workers are encouraged to view their bodies as just another type of resource to be harnessed for capitalist production (72).

Beyond ensuring candidates are giving voluntary, informed consent for enhancements and guaranteeing equal employment opportunity for those who choose not to engage with these technologies, another way to avoid the issues considered above is to reconsider how some professions are organised. It has already been noted that enhancement to stay awake for long periods of time will not represent an undue incentive in terms of potential financial benefit if workplaces are forbidden from having excessive shift lengths. I propose that to avoid the issue of perpetual growth driving more extreme working conditions, possibly facilitated by enhancement technology, similar restrictions may need to be applied in some industries regarding outcome-based measures. If an employee has engaged in cognitive enhancement to complete their work more efficiently and thereby gain more leisure time, it might be considered unfair to require them to work the same hours as a slower employee who produces far less work in the same timeframe. The enhanced individual would essentially be being punished for their efficiency with more work. This is an existing issue in workplaces that run to traditional time-based models, rather than focusing on output. For some occupations, the physical presence of the employee for the set time period is essential to the operation of the business, e.g. customer service staff need to be available any time customers are present. However, for many other industries, a structure built on equal output may be fairer for everyone, assuming the aforementioned restrictions on excessive shift length are already in place to prevent impossible standards being set for unenhanced employees by their enhanced co-workers.

Given some forms of human enhancement are already present in the workforce, our job now is to integrate them in a way that maximises their benefits while minimising potential harm, including through strengthening anti-discrimination laws and improving working conditions for all staff. It will also involve making decisions about what kinds of enhancements, if any, we might want to prevent being developed, or at least exclude from consideration in employment.

7.5.3 Protecting Authenticity of Desire

While I argue that Aldous Huxley's famous dystopian classic *Brave New World* (1932) is frequently misused in bioethics scholarship to prejudice audiences against biotechnological developments, there is one aspect of this story that is particularly relevant for this discussion. In this text, citizens are bioengineered to fulfil assigned roles in society, with their physical strength and mental capacity tailored to these positions. This is not merely a case of enhancing citizens to best achieve society's goals through eugenics, but also deliberately doing the opposite: manipulating fetal growth to limit cognitive development for those destined for low social status. The rationale behind this system is to produce a population where everyone is happy with their lot in life. At the heart of the "yuck factor" and anxieties about the supposed "unnaturalness" of transhumanism, as well as fears regarding the potential violation of the principle of justice that allowing the creation of a race of H + "haves" and H "have nots" would entail, is a unified concern that is summed up in the question: what does it truly mean to be human? This worry plays out with attempts to separate treatment and enhancement, with the latter often seen as posing a threat to some concept of what humanity should be. But while most baulk at the idea that being human can be distilled to some functional capacity, particularly given the diversity of human abilities represented in the species, I argue one potential response to this question is: freedom of desire. The people in *Brave New World* have been violated, not by abuse or oppression, but by the manipulation of their desires.

The challenge promoting authenticity of desire poses for regulating transhumanist interventions is captured well by Lilley (2012) when he states:

if we accept the accusations leveled by both sides in the debate we would be in the untenable position of "dammed if you do, dammed if you don't." The conservationists contend that if liberal democracies go down the transhumanist path and allow free choice for enhancement technologies, consumers knowingly or unknowingly will suffer modifications that diminish free will. Governments would exploit them. According to the transhumanists, if liberal democracies take up the conservationists' cause, the state would become more involved in the regulation and control of reproduction, the body, and parenting through banning enhancement technologies, and monitoring and policing illicit use. This would entail an increase in state power and loss of personal autonomy. Each side raises the specter of totalitarianism. They dispute which truly is the destructive path, but not the high stakes involved (10-11).

It is clear that if promoting liberty and autonomy is our goal, allowing interventions that can manipulate desire will undermine this project. At the very least this suggests governments and corporations should be prevented from promoting modifications that would diminish the free will of citizens or employees. This does not preclude their providing other forms of enhancement, as even if there is a vested interest in employers funding interventions to enhance their employees' concentration, for example, they cannot force the relevant individuals to use their enhanced concentration abilities to produce better work. As is often noted with genetic engineering of children, selecting embryos for musical talent in no way guarantees you will get a concert pianist in the family, as the child may choose to devote their energies and talents elsewhere. However, were parents able to select both for musical talent *and* desire, this may materially damage their offspring's right to an open future (Millum 2014).

If what makes us human is what we love and strive for, rather than the activities we can do, humanity will not be threatened by transhumanist interventions that improve functional capacity but do not interfere with authentic human desire. If protections are in place to ensure providing more choice for some people does not mean leaving fewer options for others, including in employment, the fact an augmented body may be able to bench 500 kg while a typical body cannot, need not leave anyone worse off (unless benching 500 kg is what one most authentically desires of course!). Importantly, concerns about inequality in the face of transhumanism are valid and require proactive policy changes to ensure people choosing enhancements are doing so in line with their own ideas of what makes a good life, and not due to external pressure, including workplace competition.

7.6 Conclusion

The goals of the transhumanist movement are to promote wellbeing by transcending the limitations of human biology. However, within a neoliberal capitalist society, such a goal might be co-opted by corporate interests and lead to the exploitation of both enhanced and unenhanced workers. With its core principles including promoting autonomy for individuals and justice for all members of societies, the field of bioethics is well-suited to evaluate the promises and perils of transhumanist interventions in human existence, and their potential role in the workplace of the future. Since it is likely many transhumanist interventions are inevitable, regardless of policy decisions, there is a need to consider how we want to integrate these technologies into society, capitalising on their benefits while minimising potential harms. With conscious and careful planning, transhumanist interventions can promote human flourishing through enhancing personal choice and individual expression.

References

Barns, I. 2018. Can we re-imagine a good society after Neoliberalism? *Arena Journal* (49/50): 122–168.

Beauchamp, T.L., and J.F. Childress. 2001. *Principles of biomedical ethics*, 5th ed. Oxford: Oxford University Press.

Bloom, P. 2020. *Institutions and governance in an AI world: Transhuman relations*. Switzerland: Palgrave Macmillan.

Bostrom, N. 2005. A history of transhumanist thought. *Journal of Evolution and Technology* 14 (1): 1–25.

Chambers, T. 1999. *The fiction of bioethics: Cases as literary texts*. New York: Routledge.

Clark, I., and H.P. Landolt. 2017. Coffee, caffeine, and sleep: As systematic review of epidemiological studies and randomized controlled trials. *Sleep Medicine Reviews* 31: 70–78.

Daniels, N. 2008. *Just health: Meeting health needs fairly*. Cambridge: Cambridge University Press.

Dewar, L., and R. Heuberger. 2017. The effect of acute caffeine intake on insulin sensitivity and glycemic control in people with diabetes. *Diabetes & Metabolic Syndromes: Clinical Research & Reviews* 11 (supp. 2): S631–S635.

Elshtain, J.B. 2004. The body and the quest for control. In *Is human nature obsolete? Genetics, bioengineering, and the future of the human condition*, ed. H.W. Baillie and T.K. Casey, 155–175. Cambridge: The MIT Press.

Fletcher, D.-J. 2014. Transhuman perfection: The eradication of disability through transhuman technologies. *Humana. Mente Journal of Philosophical Studies* 26: 79–94.

Fox, S. 2018. Cyborgs, robots and society: Implications for the future of society from human enhancement with in-the-body technologies. *Technologies* 6 (2): 50–61.

Grau, M. 2003. 'Redeeming the Body' in Late Western Capitalism: Pondering salvation and artificial intelligence. *Journal of Women and Religion* 19 (20): 144–156.

Huxley, A. 1932. *Brave new world*. London: Chatto and Windus.

Kass, L. 1997. The wisdom of repugnance. *The New Republic* 216 (22): 17–26.

Katz, S., and J.A. Gish. 2015. Aging in the biosocial order: Repairing time and cosmetic rejuvenation in a medical spa clinic. *Sociological Quarterly* 56: 40–61.

Kurzweil, R. 2010. *The singularity is near: When humans transcend biology*. London: Gerald Duckworth & Co.

Kwame, A., and P.M. Petrucka. 2021. A literature-based study of patient-centered care and communication in nurse-patient interactions: Barriers, facilitators, and the way forward. *BMC Nursing* 20: 158.

Le Dévédec, N. 2020. The biopolitical embodiment of work in the era of human enhancement. *Body & Society* 26 (1): 55–81.

Lilley, S. (2012). *Transhumanism and society: The social debate over human enhancement*. Netherlands: Springer.

Marchant, G.E., and A. López. 2012. The (in)feasibility of regulating enhancement. In *Building better humans: Refocusing the debate on transhumanism*, ed. H. Tirosh-Samuelson, and K.L. Mossman, 255–269. Frankfurt: Peter Lang GmbH.

Mayers, L. 2018. Sydney bio-hacker who implanted Opal Card into hand fined for not using valid ticket. *ABC News*, 16 March 2018. https://www.abc.net.au/news/2018-03-16/opal-card-implant-man-pleads-guilty-transport-offences/9555608

McNamee, M.J., and S.D. Edwards. 2006. Transhumanism, medical technology and slippery slopes. *Journal of Medical Ethics* 32 (9): 513–518.

Mill, J.S. 1859. *On liberty*. London: J. W. Parker and Son.

Millum, J. 2014. The foundation of the child's right to an open future. *Journal of Social Philosophy* 45 (4): 522–538.

More, M., and N. Vita-More. 2003. Principles of extropy, version 3.11.2003. Original version. The extroprian principles. *Extropy, 5*(5) (May 1990). www.extropy.org/principles.htm

Mossman, K.L. 2012. In sickness and in health: The (fuzzy) boundary between 'therapy' and 'enhancement. In *Building better humans: Refocusing the debate on transhumanism*, ed. H. Tirosh-Samuelson, and K.L. Mossman, 229–254. Frankfurt: Peter Lang GmbH.

Niemelä, J. 2011. What puts the 'yuck' in the yuck factor? *Bioethics* 25 (5): 267–279.

O'Carroll, R.E., C. Foster, G. McGeechan, K. Sandford, and E. Ferguson. 2011. The 'ick' factor, anticipated regret, and willingness to become an organ donor. *Health Psychology* 30 (2): 236–245.

Olenik, S., H.S. Lee, and F. Güder. 2021. The future of near-field communication-based wireless sensing. *Nature Reviews Materials* 6: 286–288.

Peters, T., K. Lebacqz, and G. Bennett. 2008. *Sacred cells: Why Christians should support stem cell research*. Lanham, U.S.A: Rowman & Littlefield Publishers, Inc.

Plows, A., and P. Boddington. 2006. Troubles with biocitizenship? *Genomics, Society and Policy* 2 (3): 115–135.

Pruchnic, J. 2016. *Rhetoric and ethics in the cybernetic age: The transhuman condition*. UK: Routledge.

Reveley, J. 2015. School-based mindfulness training and the economisation of attention: A Stieglerian view. *Educational Philosophy and Theory* 47 (8): 804–821.

Szocik, K., R. Campa, M. Boone Rappaport, and C. Corbally. 2019. Changing the paradigm on human enhancements: The special case of modifications to counter bone loss for manned Mars missions. *Space Policy* 48: 68–75.

Smith, S. 2016. 'Limbitless Solutions': The prosthetic arm, Iron Man and the science fiction of technoscience. *Medical Humanities* 42: 259–264.

Sparrow, R. 2010. Implants and ethnocide: Learning from the cochlear implant controversy. *Disability and Society* 25 (4): 455–466.

Viet, W. 2018. Cognitive enhancement and the threat of inequality. *Journal of Cognitive Enhancement* 2: 404–410.

Wilson, D. 2013. What can history do for bioethics? *Bioethics* 27 (4): 215–223.

Evie Kendal is a bioethicist and public health researcher in the School of Health Sciences and Biostatistics, Swinburne University of Technology. Evie's research interests include ethical dilemmas in emerging biotechnologies, space ethics, and public health ethics."

Chapter 8
Transhumanism—Agency Enhancement

Steven McFarlane

Abstract Hundreds of millions of dollars are spent each year on diet pills, gym memberships, self-help seminars and books, and psychiatrist visits to help quit smoking. What if we could provide more effective treatments or interventions that will help these people achieve their goals? I introduce the concept of "agential enhancement" and explore some of its contours. Agential enhancement involves enhancing the bio-chemical mechanisms associated with an agents' ability to carry out their goals. It is an enhancement of one's executive functioning. Agential enhancement might strengthen an agent's willpower in the face of temptation, allow one to regulate one's own emotions, and give one the ability to deepen one's convictions to carry through with difficult or complex tasks one has already begun. Other forms of enhancement that have received attention include physical, cognitive, mood, and moral enhancements. Agential enhancement may interact with these other types, but it is distinct. Unlike moral enhancement, agential enhancement would likely be desired by most people and would likely avoid some concerns with paternalism.

8.1 Introduction

Speaking in rough generalities, agency involves doing what you want to do when you want to do it, and in the way you intended. Something that lacks agency cannot achieve goals in the way it intends, or it may be incapable of wanting to do anything at all. Agency is like fluency in that it comes come in degrees. There is a lower threshold beneath which there is no fluency at all; above the lower-bound threshold, there is a wide range of degrees of proficiency.

We can tell when people are more or less fluent in their agency, both from the internal experience of it in ourselves, but also in observing it in others. A drunk driver has difficulty walking a straight line or reciting the alphabet backwards (illustrating the idea that impaired driving can be a matter of impaired, though not absent, agency).

S. McFarlane (✉)
Madison, United States

Sometimes we are listless and complete tasks half-heartedly; other times we are crossing items off our to-do lists at a record clip.[1]

So, agency can come in degrees and impaired agency can be problematic, as in the case of impaired driving. What if it were possible to improve agency? We already attempt to do this in many ways. We use planners, incentives (both positive and negative), self-help media, and even hypnotism to help us achieve our goals.

In this chapter, I will offer some programmatic thoughts about the possibility of *enhancing* agency. I try to rely on an intuitive notion of agency, without delving into thorny philosophical debates about what agency truly is or providing the one true analysis of the concept, as I think such debates can only be distracting in this context.[2] Instead, I will rely on what one might call "thick" agential concepts and examples, on analogy with Nancy Cartwright's work on thick causal concepts (see, for instance, her (2007)). The basic idea is that we are more directly aware of occurrences of causation in the real world than any particular philosophical theory of causation. The heater causes the water to boil. The dropped cigarette causes the forest fire. Forgetting to water the plants causes them to die. Can any single theory of causation accommodate these diverse cases? It is unclear, but that does not seem to detract from our ability to identify these occurrences as causal phenomena. Similarly, I will appeal to cases easily identifiable as agential occurrences, and treat a unifying theory as underdetermined.

As will be seen in 9.3, I rely on a psychologistic model of agency, but this is primarily for the purpose of introducing the topic, and not some final word on how we should think about the metaphysics of agency. I simply think it is natural for someone to say, "I want to lose ten pounds this year" and for us to understand the speaker to have at least one desire, and in many instances, this speaker has thought of a plan for how they will fulfill it, say by starting on a diet. But nothing important rides on psychological states being fundamental to agency, or of plans being a psychological primitive, or of certain types of mental states reducing to others. Perhaps at the end of the day, neuroscientists and endocrinologists, or from a separate perspective, sociologists, could discuss the key aspects agency without any reference to psychological states. I believe such a hypothetical state-of-affairs is compatible with my project in this chapter.

---

[1] Shepherd (2017).

[2] Pickard (2011) claims that the "common sense" notion of agency involves 1. A capacity to choose and 2. A capacity to execute on this choice (p. 212), and that many historical philosophers, from Aristotle to Kant, share this notion. Perhaps a capacity for choice is required for agency, but not all agency requires active choosing (or even active abstinence from choosing). Suppose I am standing in line at the store. I will have made many choices while shopping and going to the checkout counter, but the very action of *standing* in the line is not something I've devoted any mental activity towards whatsoever, and yet it involves agency. If I have been standing for a long time, an ache in my feet might come to my attention, and I might start looking for a chair. At that point, I might make a choice to sit or stay standing. There are rejoinders to this thought, and philosophers might go back and forth, but this type of analysis of fine details of the concept of agency is the sort of thorny issue I wish to avoid in this chapter. For a recent volume covering many aspects of the philosophy of agency see Ferrero (2022).

Perhaps more edifying than a unifying account of agency, I will discuss failures of agency to help illustrate why we ought to be interested in the possibility of enhancement. I will also place agency enhancement in context with respect to research on cognitive and moral enhancement. Last, I narrow the focus to a discussion of a particular type of agential enhancement, enhancing desire, to put more meat on the bones of this intriguing idea.

8.2 Failures of Agency

Impaired agency is not always recognized as its own category of concern, but it ought to be. Many subcategories of failures of agency are recognized as such: addiction, health issues related to the difficulties maintaining healthy diet and exercise routines, and impulse control, among others. For example, take this powerful description of gambling addiction provided via a passage in a gambling activity log as reproduced in Natasha Dow Schüll's masterful book *Addiction by Design: Machine Gambling in Las Vegas*:

3 a.m., was nearly alone, had to go to the bathroom, didn't want to leave the machine.

5 a.m., still there, choking on smoke, starving, cramping from bladder pain, butt hurting from sitting.

6 a.m., finally got up, put my coat on but still couldn't leave. Got attendant to watch machine while I peed. Almost cried with relief. Looked at myself in bathroom mirror, was shocked at what I saw. I do not ever want to look on the face of that woman again—the desperate one, the smoky, hungry one who doesn't have the sense to go to the bathroom or go home. Continued playing—standing up, coat on.

8 a.m., breakfast eaters arriving and I became terrified that someone I knew would see me. Finally left …

How did I get to this point? 15 h? I've never done anything in my life for 15 h straight, except take care of my babies. I'm well past that point in my life, could be a grandmother. And what kind of Grandma would that be? Some idiot with no self-control, who becomes paralyzed, hypnotized—by what? A machine? The music? The lights? WHAT IS IT??

I've lived a somewhat charmed life—never alcohol, never drugs, never running or being run on. Good and accomplished kids. Opportunities. Life has been sweet, wonderful, and very blessed. I don't understand this.

This gambler, pseudonym Darlene, has the markings of *failure of agency*. In an important sense, Darlene wishes she were not behaving as she is. She feels that she is not in control, and she does not understand why this is. It is not primarily a cognitive failure. She clearly judges that what she is doing is suboptimal, even harmful. She thinks it would be good to leave, she wants to leave, and yet she cannot *will* herself to leave. She describes herself as paralyzed. Though not feeling constrained or boxed in, she feels as if her internal behavioral mechanisms are not responsive to her judgment and desire to leave (or at least go to the bathroom).

Personal stories like these fit within a striking broader context. According to recent estimates, the sales for diet pills have turned into a multibillion-dollar per year industry in the U.S. alone.[3] The "Personal Development Industry"—self-help books, videos, courses, retreats, etc.—also measures in the billions of dollars per year. The profitability of commercial gyms is in large part driven by people paying for the privileges of a membership they will hardly ever use. Prospective exercisers not only buy expensive monthly or yearly memberships only to regularly skip their work-outs, they also neglect to cancel their frequently underused memberships at the end of the contract period (DellaVigna and Malmendier 2006). Other common agential maladies include road rage, drug addiction, binging video games and other various forms of escapism, and chronic procrastination. Other attempts to address these maladies, or at least evade the corresponding negative consequences, include liposuction surgery, hypnosis, herbal remedies, meditation, and, too often, various attempts at self-flagellation in the hopes of punishing one's way to successful motivation.

Failures of agency are myriad, so providing a full catalogue is virtually impossible. However, there are common types. One familiar type of agential failure is *akrasia*, or weakness of will. An agent acts akratically when he acts in a way that is contrary to what he judges would be best. Familiar examples include sacrificing a self-imposed diet by having a second dessert, blowing off a meeting with the personal trainer in order to binge-watch a full season on Netflix, blowing off studying for the important exam until pulling an all-night cram session the night before the exam, etc. In this case, the agent has an idea of what is choiceworthy but acts contrary to that judgment. A special case of *akrasia* is a bad habit, including malign addictions.[4] In this case, the agent continually acts against what they judge is best. In theory, agential enhancement would boost the agent's motivation to act as he judges best. It would be a corrective, from his own perspective.

Second, there is lack of resolve (Holton 2009). Here, an agent has made a commitment to carry out a particular act, regardless of whether he judges it best to do. Having made this commitment, the agent wavers and perhaps fails to follow through on it. Lack of resolve is distinct from *akrasia* in that, unlike the latter, the agent need not judge it best to follow through on his commitment. It is enough that he is committed. For example, suppose that on a dare from a group of peers, I agree to jump off the 30-foot-high bridge into the river. I now realize it was not wise to agree, but now I have made a commitment and intend to do it. It would be a lack of resolve and would *feel* regrettable in some way, were I to chicken out of jumping (even though I know it would be good to do!). Of course, not all commitments are contrary to better judgment. Here again, agential enhancement could work as a booster shot to motivation to follow through in executing the relevant commitment.[5]

---

[3] Austin et al. (2007).

[4] I don't mean to imply that malign addictions are mere bad habits. The relevant category is dispositional *akrasia*, and bad habits and harmful addictions belong to that category.

[5] There may be ethical concerns regarding whether one *ought* to enhance motivation or agency of this sort. I will say a bit more about the ethics of enhancement, but in this chapter, I am more concerned with exploring what agential enhancement is, what we might do with it, and how it might

A more general sort of agential enhancement is emotion regulation. Both episodic emotional outbursts (i.e. a temper tantrum or panic attack) and long-term emotional maladies (i.e. depression) can play a role in agential failure. It appears that we possess both conscious and non-conscious mechanisms for emotion regulation (Mauss et al. 2007; Gross 2015). Agential enhancement could potentially aid with either type of method. An example of conscious regulation might include counting to ten before acting out of anger. Counting to ten may help reduce incidents of aggressive behavior. Enacting this strategy isn't easy to do, and AE could in principle make it easier. Agential enhancement may aid the non-conscious mechanisms responsible for decreasing arousal and maintaining emotional allostasis.

An enormous amount of effort and money is spent in attempts to help people achieve goals that are clear in their own minds. Think of all the money thrown at medication and treatments to break addictions, dieting and associated pills and weight loss gimmicks, gym memberships, self-help books, hypnosis, etc. The list goes on. In theory, improving agency could replace many of these less effective techniques and potentially bring relief to many.

I have not provided an exhaustive list of agential failures, and I have not provided an account of how to draw boundary lines between the categories of things that are agential failures and things that are not agential failures, but I hope that the idea that people can fall short of their goals in many ways, and that these failures are sometimes due to "internal" failures of control—times when people someone need only walk away from the slot machine, or roll out of bed, or refrain from eating more cake is intuitive. With a primarily ostensive conception of agential failure in mind, we can turn to what it might look like to enhance agency via biochemical intervention.

8.3 Enhancing Agency

Consider some methods of biochemical interventions with which we are already acquainted. Athletes use steroids and other performance-enhancing drugs (PED's) in sports like bodybuilding. The use of this sort of biotechnology is by now familiar, and not a matter of science fiction. People regularly use advances in our knowledge of the human body to make intentional alterations to what is thought of as normal bodily development. Similarly, recent estimates show over two million children and adolescents used medication to treat ADHD (attention-deficit hyperactivity disorder) in recent years (Danielson et al. 2018). Many of these children and adolescents use this medication to change their biochemistry in large part to do better at school. Clearly, there are different reasons and justifications for these types of interventions than there are for bodybuilders using PED's. Still, we can view use of Ritalin as a familiar case of using biotech to enhance performance of a different kind—in the classroom rather than at the gym.

---

work rather than whether we ought to do it. This is more of a diagnostic treatment, rather than a prescription.

Philosophers have been discussing the uses of biotechnology and have names for the kinds of uses I mention above. Using PED's is called physical enhancement. Using medications to improve our ability to think and learn is called cognitive enhancement. Selective serotonin reuptake inhibitors (SSRI's), a common pharmacological type of anti-depressant, might qualify as mood enhancement technology. The two primary ideas behind enhancement biotechnology are (1) our bodies are biological systems we can manipulate directly through technological means, just like any other natural system, and (2) we can manipulate human biological systems so that they can perform better than if there were no technological intervention.

Bioenhancement is often contrasted with other forms of improvement. For instance, by regularly lifting weights and eating protein, one can become stronger and more muscular than average. Regular exercise would not count as bioenhancement as it is typically understood in the philosophical literature on the topic. Similarly, studying for a long period of time does not count as cognitive enhancement, and so on. We may have reason to question some of these suppositions, but first it is important to situate the commonly understood presumptions in debates over bioenhancement.

What sets bioenhancement apart from other forms of improvement is the means of intervention. Bioenhancement by-passes typical processes of development or growth—be they psychological, social, hormonal, neuronal, or the like—and instead directly intervenes on the relevant biological level, whether it be particular organs, neurons, processes, or systems. I will provide other examples to clarify the contrast between enhancement and typical development as we go, but keeping PED's in mind is a good place to start. Athletes who use PED's gain more muscle more quickly than they would otherwise, and some athletes use PED's so effectively that virtually no one could replicate their success in muscle-building without using PED's. In other words, natural bodybuilders, no matter how much they exercise or how much protein they eat, could not reach the heights of the best bodybuilders we currently have. In this case, lifting weights is something of a traditional form of improvement; PED's qualify as enhancement.

To get a better picture of what it means to enhance agency, specifically, it is useful to consider a categorization of types of psychological attitudes and their functions, as enhancing agency is best understood as enhancing our ability to execute on our plans. Table 8.1 includes a set of psychological attitude types, their functions, and common-sense examples.[6]

Agency enhancement might target conation, affect, or (inclusive) executive function directly, rather than cognition. (In Sect. 8.8, I discuss conation and desires in more detail). In less technical terms, agency enhancement would aim at aligning our emotions, desires, and intentions in service to achieving our own selected goals.[7]

---

[6] The table is not exhaustive, and some researchers debate whether one or more of these functions and types reduce to others. I do not offer a full defense for the metaphysics of these categories. Rather, I provide the table to better understand how agential enhancement might work. It provides a framework for understanding the *aim* of an agential enhancement intervention.

[7] In presenting these ideas, I am frequently asked about enhancing agency in the selection of goals, apart from their implementation. This seems to me to be a rich vein of inquiry, but I do not say

Table 8.1 Attitudinal function, type, description and examples

Attitudinal function name	Paradigmatic attitude type	Attitudinal function description	Examples
Cognition	Belief	What one judges to be true	Belief that $2 + 2 = 4$ Perceptual belief that grass is green
Conation (Motivation)	Desire	What one is motivated to pursue	Appetite for dessert Desire to graduate college
Affect	Emotion	Feelings or moods	Temporary affect: Joy, Envy Moods: Optimism, Depression
Executive Attitude	Intention	A plan to carry out	Intention to go to the library to return books and the bank to deposit a check, followed by returning home and completing assignment

Agency enhancement might enhance executive function, desire production and maintenance, and emotion regulation. It would aid our ability to stick with our goals in the face of obstacles, whether by boosting our motivation {enhancing desires) or by helping us resist temptations or distractions {by enhancing executive function or ability to regulate emotion).

8.4 Enhancement Versus Therapy?

There is another common distinction to address—the distinction between therapy and enhancement. Therapy might be defined roughly as an intervention on a patient's behalf that is a remedy for an illness, injury, or disability. For instance, doctors sometimes prescribe anabolic steroids to patients with AIDS to help maintain muscle mass and overall weight. Enhancements, on the other hand, are roughly defined as unnecessary interventions on a patient's behalf to improve beyond normal functioning or capacities. Intuitively, there is a contrast between undergoing surgery to remove potentially harmful wisdom teeth (therapy) and merely having one's teeth whitened every six months (enhancement). In short, it seems that therapy gets you back to normal; enhancement puts you above normal. Writing about what I am calling the therapy/enhancement distinction, Dees (2007) claims:

The distinction might, for example, be important for deciding for which treatments insurance and governments will pay. Because some uses of the drugs do something

---

much about it here. I focus primarily on instrumental uses of enhancement and agency in order to avoid larger, more intractable issues for a first attempt at articulating a concept.

more than restore a patient to "normal" functioning, we can say that in those cases the drugs are not medically necessary and so we should not expect government or insurance to pay for them. This standard is, after all, is [sic] the one used for cosmetic plastic surgery. But in cosmetic surgery, the principle of autonomy rules: patients can decide for themselves what is excessive and what is good (p. 377).

As Dees notes, there is precedent for making such a distinction, and many people intuitively grasp a more urgent need for people to have access to therapy, but not necessarily for enhancement.

However, as others have noted (Bostrom and Roache 2007; Erler 2017), the intuitive distinction is surprisingly difficult to support analytically. One question regards how to determine what qualifies as "normal" functioning or status. There are at least two cross-cutting types of reference frames for what counts as normal. First, we might look at what is normal for a particular individual versus what is normally distributed among a population. Second, we might look at what is normal relative to age versus what is normal across the lifespan (Bostrom and Roache 2007). I consider these options in order.

The case for relativizing to an individual rests on the thought that each of us has individual capacities with natural biological limits. Phenotypic variation across a species is no secret, and we have every reason to suppose such variation extends to individual differences in capacities for strength, height, weight, intellect, resistance to disease, or self-control. Because there are degrees of variation in these traits, an individual's "normal" capacity may be hindered or damaged but *remain atypical relative to the rest of the population*. Bostrom and Roache ask us to consider a person who "might have a recognizable neurological disease that reduces her cognitive capacity by one standard deviation ($1\ \sigma$), yet she would remain above average if she started off $2\ \sigma$ above the average" (p. 121). Suppose a doctor could cure this person's disease. Would this qualify as a therapy or enhancement? If we compare to what is normal for this person, the treatment counts as therapy. However, relative to the larger population, intervening would cause this person's cognitive capacity to increase her cognitive atypicality even further beyond what is normal, and hence look like something closer to enhancement. What is ordinary for one person might be extraordinary for another.

The fact that it would be strange to say that restoring someone with atypically high cognitive capacity to a previous normal for that person counts as "enhancement" is not in itself an argument against relativizing the therapy/enhancement distinction to individual personal capacities. But, in practice, we just typically do not think of the distinction in this way. Major league sports associations ban PED's from all players, not specific individuals relative to their individual capacities. Likewise, only those with ADHD or a similar condition are permitted to receive prescriptions for Adderall or Ritalin.

This might lead one to think that we can base a therapy/enhancement distinction on whether a patient has contracted a specifiable disease or symptom, regardless of what one's "normal" capacity lacking that symptom is. A problem for this view is that how we define diseases or negative conditions is often, though probably not always, contextual and based on typicality. Allen Buchanan provides a useful example: vision and visual impairment (Buchanan 2017). There is nothing partic-ularly special about 20/20 vision. Someone whose vision is worse can, these days, cheaply and easily procure remedial vision technology: eyeglasses, contact lenses, or laser surgery. Contact lenses, as prescribed by an optometrist, will typically get one's vision back to 20/20. In truth, we have the technology to give people vision better than 20/20. (Buchanan had eye surgery that improved one eye to be better than 20/20. See Buchanan 2017, p. 8.) Why do we view 20/20 as the appropriate default for vision, rather than 10/10, or even 30/30? After all, it's not hard to imagine that at some point in the history of *Homo Sapiens*, average eyesight was not 20/20. It is also easy to imagine a future where the average eyesight is not 20/20.

The Snellen ratio, the typical values of which are 20/20, is a normalized ratio based on the ability to discern letters on a chart of increasingly smaller size at a distance of twenty feet. The American Optometric Association explains that "If you have 20/20 vision, you can see clearly at 20 feet what *should normally* be seen at that distance. If you have 20/100 vision, it means that you must be as close as 20 feet to see what a person with *normal vision* can see at 100 feet" (emphases added).[8] Note that the sense of 'normal' here is something of an educated guess of statistical average.[9] Having less than 20/20 vision is not really a disease or deficiency unless we agree 20/20 is what the default should be.

Similar factors are relevant in the case of height, IQ, athleticism, capacity for memory, and perhaps many other attributes. I do not mean to imply that all diseases or negative conditions are contextual in this way. Some diseases are clearly detrimental regardless of averages or typicality. However, the negative conditions of some types of disease or disabling conditions are, at least as commonly understood, contextual in this way. So, it seems we need to look at the prospects for the view that the therapy/enhancement distinction is grounded in the traits of distributed populations, rather than focusing solely on the traits of an individual.

One of the primary difficulties for the view that we ought to use what is normal for a population as the dividing line between therapy and enhancement is determining the correct relevance criteria for defining the appropriate population. It makes little sense to define what is normal by averaging the traits of the entire world population. What would a normal diet be? Which medications would be seen as normal? For some traits, it would be easy to find an average, but we would be missing important context. For instance, it would be relatively easy to determine the average weight of a global citizen. But this information would be useless or worse if we do not consider

---

[8] https://www.aoa.org/patients-and-public/eye-and-vision-problems/glossary-of-eye-and-vision-conditions/visual-acuity.

[9] The history of the Snellen Ratio reveals a fair amount of arbitrariness. Dutch ophthalmologist Hermann Snellen, looking for a more standardized test to measure visual acuity, created the now familiar Snellen Chart ('E' written in large font on top row, above five rows of letter of diminishing font size) because of its more simple but still rigorous mathematical utility. Why use these particular font sizes? Why is twenty feet the relevant distance, rather than ten or thirty? In large part, the answer is convenience. There is essentially no reason ophthalmologists could not use a test with fonts at such and such a size at thirty feet.

many other factors, such as age, sex, height, and so on. So, we need to limit the population in reasonable ways.

There is not much hope in finding good criteria. The first place to look might be geographic boundaries. But what about shared ancestry among a people with variable geographic disbursement? Which demographic features are relevant? Age will surely factor in greatly. We know that educational opportunities are relevant to IQ scores (Ritchie and Tucker-Drob 2018). If biology, developmental stage, geography, social environment matter, then they all must factor into the relevant comparison class.

It will be difficult to find appropriate criteria. In particular, it will be difficult to find them without begging any questions. These problems for supporting a therapy/enhancement distinction are well-known, and I have been convinced by these considerations on their own that there are significant problems for trying to draw an important normative dividing line between enhancement and therapy. But I think another reference frame leads to a fatal epistemic problem, one that is irresolvable at the current time. To start, note that we often treat contemporary life as though it were not a step in time, but as if the status quo will continue roughly as it is. In fact, while there is a great deal of diversity across the globe now, there is unimaginably more possible variation if one takes into account humanity's past and future. Take height as an example. We know that what is considered normal height in the 2000s varies greatly correlating with geography. But, if we were to travel in the past a few hundred years, the average U.S. male would be seen as tall in most if not all societies at the time (Similar remarks could apply to many athletic and cognitive capacities).

Comparisons with past humans in determining what is typical or average are a problem, but not necessarily irresolvable were researchers to somehow discover relevant data regarding vast numbers of past humans. But matters are much worse when we turn our attention to what are "average" or "normal" human capacities in the future, or to what might qualify as a disease or disability appropriate for therapy. Will dying at 70 be viewed as tragically early and often preventable if the human lifespan reaches 140 years, just as a death at 40 seems now? Will the Flynn Effect (Trahan et al. 2014), the trend wherein the normalized average IQ among Americans increases by roughly 3 points per decade, continue apace, ensuring that future generations have much higher IQ's than we do now? Or might the trend reverse?

Given that we do not know what the trends and limits for human strength, intelligence, or lifespan are, it is fair to conclude that we cannot now determine the point at which an intervention counts as therapeutic or enhancing on the basis of finding some "normal" relative to a comparison class. To summarize the preceding considerations in argumentative form:

In order to know where to draw the line between therapy and enhancement, we must know what is "normal" or "average."

In order to determine what is normal or average for the traits relevant to the therapy/enhancement debate, we need to know what the distribution of these traits are relative to a determinate population (or we need to know of a representative population from which we draw trustworthy extrapolations).

We have no good way of identifying what the determinate population is or could be, as this has changed in the past and could (is likely to) change in the future and any attempt to provide a representative subset will be missing relevant data.

Without a viable comparison class, we cannot determine what is normal or average.

Therefore, we cannot know where to draw the line between therapy and enhancement.

This argument presumes we are using a non-normative sense of "normal." In theory, we could someday know where to draw a line, but not now.

It is possible to temporally index the sense of "normal" one wants to use in drawing the therapy/enhancement distinction. One can look at what is normal *now* for a person with this ancestry, in this country, at this age, and so on. But this looks like special pleading for an allegedly morally important quasi-metaphysical distinction. If there is an important distinction between therapy and enhancement that we morally ought to recognize in our practices, then it would appear that the distinction ought not be attributable to pragmatic considerations in the face of uncertainties. If we ought not provide cognitive boost to an American in the year 2000 C.E. with a normalized IQ score of 100 because that would mark it as enhancement, what makes American in the year 2000 C.E. special? Suppose that in the year 2060 C.E. the normalized IQ score of 100 is equivalent to 80 in the year 2000. Which is the relevant comparison class? What non-artificial reason would we have for selecting 2000 C.E.? If the answer is that we do not know what the normalized IQ score of an American in 2060 C.E. will be, I concede that point. But I think it carries the implication that we simply do not know what the relevant comparison class ought to be, not that we can simply ignore the issue.

I have been treating "normal" as something like the arithmetic mean, but a fair objection might be that this is not what we always mean in ordinary parlance when we use the word 'average' or 'normal'. "The average person has two ears" is something one might say that many would regard as true, but is false as a claim regarding the statistical mean of all persons. But hopefully, it is clear that whatever descriptive, statistical measure we use: median, mode, standard deviation variances, etc., we would be extrapolating into a great known unknown were we to assume we know what the biology of the humans of the future will be like given what we know of the past and present.

One might wonder whether looking at statistical notions, whether for an individual or population, is the wrong starting place. Perhaps we should not look at typicality in determining what is "normal," but instead look at "normal functioning" or "normal development," where the notion of "normal" is tied to something like teleology or proper functioning. Or perhaps we ought to do away with the assumption that "normal" functioning or capacity be understood descriptively, but understand "normal" as a normatively loaded category? In either of these cases, it must be demonstrated that the relevant conception of what is normal does not beg important questions. There is the potential for committing a version of the naturalistic fallacy. Suppose that my genetic inheritance, in combination with my developmental history and current environment, conspire to make me such that I produce such and such "normal" levels of testosterone. If I can safely increase my testosterone level through some intervention, why would the fact that I am bucking my genetic inheritance or developmental history matter? In what is their normative authority grounded? These views deserve more attention, but I will move on acknowledging that questions regarding the normative force of these views is subject to continuing debate.

Some argue that though the distinction between therapy and enhancement is vague, that does not mean we cannot tell the difference when we see clear cases (Dees 2007; Wiseman 2016). Recall the case of steroids for a patient with AIDS versus a competitive bodybuilder. As in other cases of vagueness, this would allow for problematic cases sandwiched between clear cases. Just because I cannot tell at which exact shade on the color wheel we pass from green to blue does not mean that I cannot tell the difference between clear cases of sky blue and emerald green (Dees also appeals to color p. 377). Perhaps this is correct, but if so, there is an important disanalogy between the therapy/enhancement distinction and indiscriminability of shades on a spectrum. It is not the case that there are a range of cases where it is difficult to tell whether they are normal and that reasonable people could draw the line arbitrarily. The cases we have looked at show that the trouble comes in defining relative comparison classes for normalcy in the first place, not in the indiscriminability of cases spreading outward from normal to "above" or "below" normal.

8.5   Enhancement as Improvement

The therapy/enhancement distinction lacks clarity, but the notion of enhancement is useful and important, so we are left with a puzzle: how to talk about enhancement *qua* enhancement? My proposal is to treat enhancement as an improvement by treatment. This improvement will be relativized to the individual's condition prior to receiving the treatment.

Understanding enhancement as improvement allows for a useful connection to the ersatz (in my view) therapy/enhancement distinction. Some improvements will clearly be more important than others, corresponding to worse initial conditions. For instance, patients with HIV might benefit from anabolic steroids to maintain muscle and body mass. Major league baseball players might also benefit, but since their initial condition with respect to health and the ability to retain muscle and body mass is much better, it is clearly far less important that they receive improvement interventions. In the case of scarce resources, or in decisions or policies regarding how to distribute resources, it is, of course, important to direct them where they are most needed. Attributions of instances of therapies versus enhancements might then be useful heuristics for signaling the importance of intervention. Thus, we can provide at least a partial response to Dees's concern that we need some criterion for deciding which medical treatments insurance or the government should cover.

I do not claim that understanding enhancement as improvement relative to an individual's condition captures all intuitions in the debates over enhancement. Some argue that all enhancements are impermissible, or at least that there are intrinsic *pro tanto* reasons against enhancements (Sandel 2009). If we are to understand enhancement as improvement, there are no such reasons against, only a lack of reasons in favor, relative to other uses of the technology.[10] I am not sure there is a way to square my understanding of enhancement as improvement with other accounts, so I am willing to grant that enhancement as improvement is revisionist.[11]

I believe that whatever theoretical costs that might be associated with using a revisionist account are worth it. First, as is hopefully apparent, there was never a clear consensus on what enhancement is in the first place. Second, authors working on questions of enhancement provide examples that seem to fit with improvement already. For example, with respect to various kinds of enhancements, Buchanan (2017, p. 9) claims, "Literacy is a fantastic cognitive enhancement. Being able to read and write greatly enhances what the brain can do..." In discussing moral enhancement, Harris Wiseman endorses the view that while vague and unhelpful for making a moral demarcation, the therapy/enhancement distinction has use for "descriptive purposes" and using it "as some pragmatic distinction is perfectly defensible" (Wiseman 2016). Later on, in the same chapter, he writes that "[I]n a world where moral enhancement were a reality, it might actually be better for some interventions to be compulsory. A humane intervention to constrain predatory and sadistic child rapists would be...more responsibly set forward as compulsory" (p. 202). John Harris (2007, p. 8) writes:

Many of us are already enhanced (do you wear glasses, for example?) and all of us without exception have benefitted from enhancing technologies. (For example, have you been immunized? And even if you haven't, you will have benefitted from the so-called "herd-immunity" created by the fact that others have.)

Illiteracy, predatory and sadistic impulses, or vision impairment would all typically count as deficits in need of remediation, relative to a relevant comparison class, and certainly not "normal" or typical. Thus interventions to improve them ought to count as therapy according to the traditional account. I know of no understanding of a therapy/enhancement dividing line according to which wearing glasses of the sort that many of us have (rather than some sci-fi tech) would fall on the enhancement side. And yet these authors on different sides of the debate occasionally slide into representing them as enhancements. These authors have particular points they are making with these examples, and I am taking them a bit out of context. The reason for this is not to indict these writers of inconsistency. Rather, I am marshalling support for the view that it is not altogether unnatural to understand enhancement as improvement.

A third, and perhaps the most important, justification for using a notion of enhancement as improvement relative to an individual's initial condition is that I want to discuss deficiencies that are common but difficult to diagnose as particularly in need

---

[10] There could be many extrinsic reasons against enhancement as improvement—resources could be scarce or there may be issues of fairness or equal access to treatments.

[11] Another possibly counterintuitive consequence of this characterization of enhancement is that it seems to entail that something within the realm of typical behaviors such as eating a healthier diet could count as enhancement. Personally, I do not find this counterintuitive, but it certainly is not something for which one would be penalized in a major league sport, nor does it carry any moralized connotations as other enhancements often do.

of therapy. Darlene, who has a gambling addiction, might appropriately be said to appear to be at "below normal" levels of self-control, and thus might appear to be on the therapy side of the divide. But should we consider writer's block a psychological malady? Is chronic laziness a real condition, and if so, how do we diagnose it? What about someone who overeats, but not beyond a standard deviation outside the normal variance for someone with their height, metabolism, etc.? I am not aware of clear answers to these questions, but I do not think we should hold any discussion of agency enhancement hostage while we adjudicate them. Darlene, frustrated authors and the slothful can conceivably have their respective agency enhanced, and 3it is worth considering what this might look like.

One last note on enhancement as improvement. Note that on this view, improvement is relativized to both the individual *and to the relevant dimension to be improved*. I am not arguing that enhancement entails improvement to the patient's overall condition. We can regret getting what we want, and sometimes we can get too much of a good thing. An ambitious parent might devote so much time to advancing at work that he misses the important times with his children, and look back with regret. We know this when it comes to PED's in athletics. Using steroids enhances one's ability to gain muscle, but it may deteriorate one's overall health, often damaging the heart, liver, and kidneys. It is plausible that a heavy steroid user is enhanced while becoming worse off overall. I do not think we ought to conclude that, therefore, the steroid user is not enhanced after all. Rather, we should understand that enhancement can have many different effects, some of which may make the patient worse off. Enhancement is always relativized to particular conditions, and not the enhanced individual's overall state.

8.6 Distinguishing Agential Enhancement from Other Types of Enhancement

8.6.1 Distinguish Agential Enhancement from Cognitive Enhancement

Though both agential and cognitive enhancement improve mental capacities, these capacities are defined by different functions. Cognitive enhancements help agents improve memory, form correct beliefs, or make better inferences. Agential enhancements improve agents' ability to execute in service to their goals, however they reason to arrive at the idea that they ought to pursue these goals in the first place.

It's possible to apply agential enhancement in service of goals typical of cognitive enhancement and vice versa. Undergoing AE could help one study intensely and pass up distractions, thus contributing to cognition-related goals. Undergoing cognitive enhancement may aid in achieving practical goals, such as a student's desire to achieve a perfect GPA. These points do not contradict the idea that these forms of enhancement are distinct, as the effects in question are secondary offshoots that are contingent upon the goals of the agent. Cognitive enhancement has a necessary connection to the goal of enhancing cognition, and AE has a necessary connection to executing goal states.

8.6.2 Distinguish Agential Enhancement from Moral Enhancement

Agential enhancement is also distinct from moral enhancement. Agential enhancement is subjective and instrumentally oriented. It helps an agent carry out actions to achieve his goals. These goals may be morally virtuous or vicious, however. Agential enhancement is morally neutral, and so cannot be viewed as moral enhancement.

It may well be that enhancing a vicious person's agency grants this agent with a greater ability to carry out harmful plans. A good cat burglar may wish to be the best cat burglar. But becoming the best cat burglar involves a lot of mental and physical training. This professional could undertake agency enhancement in service to his goals, and yet in no sense could this be viewed as moral enhancement.

I take it that an enhancement in service to a straightforwardly vicious end is conceptually incompatible with respect to moral enhancement. Opponents of the viability of moral enhancement (Wiseman 2016) sometimes argue that attempts to bio-engineer moral improvements may under certain conditions lead to bad consequences or increase the likelihood of moral failures. Perhaps too much empathy can be a bad thing, or perhaps a sense of justice can be warped into a destructive force. Thomas Douglas (2008), a proponent of moral enhancement, notes that it may be better to focus on *sets* of moral motivations, as it may be the case that distinct situations call for distinct motivations; a high degree of empathy may be appropriate in some cases, but not others. He argues for this reason we look for sets of motivations which can typically be expected to work well.

Regardless of the merits of these arguments, I take it that it is flatly conceptually impossible for moral enhancement to enhance a vicious agent's viciousness, but this is a distinct possibility with respect to agential enhancement. A ruthless mercenary might wish to overcome irregular pangs of conscience, as it is bad for business. Conceivably, he could undergo AE to achieve this goal. But it is difficult to see how this could possibly be a case of ME.[12] Thus, agential enhancement is a distinct concept.

---

[12] One may wish to push back from a consequentialist perspective. Suppose circumstances were such that hiring ruthless mercenaries helps the more just army win the battle, and that therefore, in the overall scheme of things, being more ruthless leads to the best state-of-affairs, morally speaking. I still think it's better if the mercenary is ruthless only to the extent he needs to be, but is capable of pangs of conscience should the need arise.

8.7 Can Agency Be Enhanced? Ought We Spend Time Thinking About It?

In earlier sections, I noted the tremendous variety of goals a population of individual agents might respectively have, and even more ways to achieve them. Some goals, such as graduating from college, are abstract and complex, requiring a vast amount of subsidiary goal attainment, e.g., waking up for 8 a.m. class. There will not be a single operation or procedure which could enhance agency with respect to all goals in the foreseeable future, and perhaps it is an impossibility.

The incredible variety and complexity of goal-oriented agency in contemporary society might reasonably lead one to wonder about the viability of enhancing agency, in any degree of generality. There are at least two distinct though related sources of skepticism.

First, one might think that agency is relegated, either entirely or at least to a great extent, to the psychological level, and thus biological interventions are unsuited to the task of enhancing agency. Thus, we are better served by focusing on the relevant psychological mechanisms directly, rather than biology.[13]

The second is that, as mentioned previously, success in achieving goals extends beyond an agent, and extends to the agent's environment, whether that be techno-logical, social, or natural. Given the prevalent role of features external to the agent in determining his success, one might be tempted to doubt that interventions on the *agent* alone, in isolation are sufficient to address agential failures at anything like the prevalence we find in society. The objection here may be that this notion of agential enhancement is too *individualistic*, in terms not just of the social, but to the broader environment in which the agent operates. For example, the common trope that if one works hard, one can make it in America is often met with eyerolls from some audiences. The reply might be that there are systemic barriers to success that one individual actor can overcome alone, or if she does, then she will have need good luck along the way.

It is true that there are important limits to what effects we can expect manipulation at the biological level to have with respect to agency. Still, there are reasons to think and talk about agency enhancement. Consider the following:

However complex the story, it remains likely the case that the psychological supervenes on the biological. Because this is likely the case, there will be some important connection between manipulations of our neural anatomy that in some way influences our psychologies. This does not by itself establish that agential bioenhancement is possible. Perhaps the biological story is impossibly complex, or unstable, or too fragile for appropriately beneficial manipulations. Perhaps the biology is comprehensible, but the relation to our psychologies remains a mystery.

While agency enhancement as some sort of general faculty may seem to some like a long shot or outright impossible (personally, I am a *bit* optimistic on this

---

[13] Zarpentine (2013) makes similar arguments regarding moral enhancement, claiming that traditional methods targeting psychological mechanisms are more efficacious than biological enhancement.

front), domain-specific agency enhancement seems eminently possible. The many researchers working on drug addiction, self-help books, overcoming social anxiety, dieting techniques, and so on are betting on this premise, as well. While there may be individuals out there who are bad agents all around, the more common scenario is for an agent to have a few weaknesses. For some, it is drugs, sometimes one particular drug. For others, it is unhealthy social attachments. For others, the issue is overeating. The biological story relating to these types of cases may prove more manageable to understand and manipulate through biotechnology. In that case, agency enhancement is still worth discussing.

Corresponding to 2., there are most likely enhanced agents among us already. For instance, Levy et al. (2014) provide good reasons to think that arguments against the moral permissibility of potential biological enhancements, as though these are a possibility in the future, not yet realized, are off the mark. Many of us are already modified, and it is plausible that at least some of us are in some sense biologically enhanced. There is perhaps too much handwringing about potential slippery slopes, and not enough examination of where we currently are on the slope. We have almost certainly descended a little already.

With respect to agential enhancement in particular, Kjærsgaard (2015), building on work by psychologists Ilieva and Farah (2013), suggests that Adderall, often connected with cognitive enhancement, is actually better categorized as "motivation enhancement." This is because empirical research suggests that participants who take Adderal report "'feeling up,' 'drivenness,' 'interestedness,' and 'enjoyment'" and that "'interestedness and 'drivenness' are related to motivational effects (rather than emotional states). The same goes for the increased levels of energy ('feeling up')" (p. 6).

Lastly, remember, we are discussing enhancement, not magic. Injecting trenbolone, a powerful anabolic steroid typically used on livestock and sometimes used by weightlifters, will not lead directly to the ability to succeed at weightlifting—one does not receive an injection and immediately have the capacity to lift a large amount of weight. Nor will steroids make one stronger in all contexts or environments. Instead, one must engage in muscle-building exercises, be free of injury, have heavy weights within one's vicinity, and so on. Nonetheless, steroids are a form of physical enhancement (cf. Buchanan makes similar points p. 163).

Perhaps agential enhancement could work analogously, functionally speaking. Suppose for sake of argument, that the ego-depletion theory of willpower is true (Baumeister 2002).[14] According to this theory, there is a limited reservoir of willpower that can be drained and refilled over time. Mentally taxing tasks drain the ego more quickly. Suppose (what at this time appears to be a counterfactual) that the neural correlates of the willpower reservoir are identified, and pharmaceutical testing reveals a drug that rapidly refills the reservoir (or retards drainage, etc.). This centralized hub of agency might then be enhanced, allowing for the agent to achieve more goals *if he engages in goal pursuit*. The agent will then be able to accomplish

---

[14] Among philosophers, Richard Holton (2009) developed a framework for human agency employing the ego-depletion model.

more mentally taxing tasks, and thereby achieve more goals. But, in the absence of such pursuit, the agent would not appear to have improved. Just as it is a mistake to say that steroids are not a type of enhancement when a couch potato who uses them does not grow stronger, it would be a mistake to think someone who does not achieve all of their goals has not been enhanced.

8.8 Case Study: Enhancing Desire

One might reasonably question what agency enhancement might look like in more detail than the broad overview canvassed so far. There are limits to what can be known about the possibilities of enhancement of course, but in this section, I will try to make the possibilities for agency enhancement more concrete by examining the possibilities for enhancing desire. I discuss desires because they are one of the central elements of agency and because it is plausible that failures of desire lead to the failures of agency I mentioned in Section II. For example, a desire to stay in bed might overwhelm the desire to get up and exercise for thing in the morning. First, I will discuss what properties we might want desires to have to see what it might mean for desires to be enhanced. Then I briefly discuss how enhancing desire might work via influencing reward prediction error and learning.

8.8.1 Desirable Properties of Desires

We want to generate motivation where we have none, but judge that it would be good to be motivated. For instance, a teacher might want to put off grading till tomorrow but be able to accurately judge that putting off grading will mean less time for research, and so judge that it is best to grade today.

We also want to maintain motivation once it has been initiated. Often, this means walling off distractions. Some typical distractions are *competing* desires, though there are other kinds of distractions. Competing desires can be particularly pernicious and all too common. I may want to quit grading and go watch the ball game. But, if I am engrossed in what I am doing, such a distraction might not arise in the first place.

Speaking of competing desires, while a completely harmonious set of desires and meta-desires is probably impossible (and that's probably a good thing), there can be value in having coherence among our desires. Here the work of Harry Frankfurt on wanting something wholeheartedly is relevant. I want to enact on the desire of my choice—not just act upon whatever desire is strongest at a given time. I also want my desires to align diachronically—I might want to act upon my strongest desire later, which might require acting on a different weaker desire now. As a professor, I may have a particularly strong desire now, which causes me to want to have a desire to grade today, though I do not now feel like grading. Bringing these desires into coherence with each other is valuable.

A fourth attractive property of desires is a sort of optionality—an ability to cease desiring, or at least opt out of acting upon this occurrent desire. It may be in the nature of desires that we cannot completely control their onset or duration; perhaps we could not ever extinguish them at will. But, to the extent that we control which of our desires we act upon out of what is typically a possible plenitude, and *also* able to stop acting on those desires, we are doing better in exercising agency. A powerful, wholehearted desire to continue grading even after the current batch of papers is finished is unwelcome. More generally, bad habits, compulsion, and addiction are cases of desires being *too* strong; they lack optionality.

To summarize, we want desires to be responsive to judgment, to be effective even in the face of obstacles and distractions, and part of a larger coherent set of desires and meta-desires. But, we want to be able to switch activities when we so choose, and so desires cannot be so overpowering as to be compulsory.

8.8.2 Multidimensional Desire Enhancement

We can now see that stronger desires are not always better. But, strengthening them is sometimes a good idea. Strengthening desires can help inoculate agents from the distraction of competing desires, or from succumbing to enervating circumstances. However, if other desires could be suppressed, then one might not need to strengthen one's primary desire relative to the others in the first place. The point is that desire enhancement will be complicated, and not a single dial controlling desire strength.

While I cannot provide a full catalogue of strategies for biotechnological methods for enhancing desires, I will offer some intriguing possibilities, relating in particular to dopamine, its role in reinforcement learning, and its connection to desire.

Current neuroscience suggests that dopamine is transmitted to the pre-frontal cortex when one's actions produce an outcome better than expected. Thus, in the simplest cases, the action tendency becomes more salient, and the agent is more likely to want to do it again in the future. Note that this is true as evaluated by the parts of the brain involved in prediction and reward signaling (ex. the basal ganglia and ventral striatum), and not necessarily the agent as a person. Snorting cocaine is often associated with a sharp increase of dopamine, and thus we are well on our way to a particularly strong action tendency formation (in this case, an addiction). Suppose that one could take a pill that, when ingested, would neutralize dopamine transmission for a short period. During that time, one could perform actions that one would *not* want to become a habit, and the action tendency will not likely take hold. Possible applications would not need to involve drugs, though in my mind this is already an important avenue of research. It could be used in relation to video games, gambling, social media use, or other activities that stop people from achieving their goals.

Of course, we could pair a pill that increases dopamine transmission for healthy actions. Suppose one needs to improve their diet but hates the taste of kale. Take a pill, wait an hour, and eat some kale. Though one may one find the taste unappealing, one may still get a strong motivation boost for the action tendency of eating kale. Our well-meaning but dawdling professor can pair a dopamine-enhancing pill with grading and be well on his way to making time for research tomorrow.

There is more to say here, including the limitations of pairing dopamine enhancement/neutralizing pills with action types. Such attempts at enhancement could be abused or lead to unintended consequences. But, surely there are limitations to current attempts of psychological interventions and words of encouragement or discouragement. It is worth exploring new avenues for enhancing agency.

8.9   Conclusion

I have attempted to provide an initial, substantive characterization of agential enhancement, justified why the concept might be of interest, and provided a possible application in terms of desires. One chapter alone leaves many avenues unexplored. Further questions relate to the social nature of agency and agents' environments, hard questions regarding the biological bases of agency, and the embeddedness and scope of goal-directed behavior (i.e. most macro-goals are distant in terms of time and in terms of the number of intermediate tasks that require completion before achieving them).

Perhaps most noticeably, I have barely skimmed the surface of the larger normative and moral questions surrounding this topic. Apart from questioning the basis for an important therapy/enhancement distinction, I have not addressed the question of whether agency enhancement is a good idea. I think it is important to have an understanding of the conceptual mechanics, and so that is what I have focused on here. I believe there will be an ongoing moral debate with respect to agency enhancement and other types of bio-enhancements, as well.

References

Austin, S.B., K. Yu, S.H. Liu, F. Dong, and N. Tefft. 2017. Household expenditures on dietary supplements sold for weight loss, muscle building, and sexual function: Disproportionate burden by gender and income. *Preventive Medicine Reports* 6: 236–241. https://doi.org/10.1016/j.pmedr.2017.03.016.

Baumeister, R. 2002. Ego depletion and Self-control failure: An energy model of the self's executive function. *Self and Identity* 1 (2): 129–136. https://doi.org/10.1080/152988602317319302.

Bostrom, Nick. and Rebecca Roache. 2007. Ethical issues in human enhancement. In *New waves in applied ethics*, ed. J. Ryberg, T. Petersen and C. Wolf, 120–152. Palgrave-Macmillan.

Buchanan, Allen. 2017. *Better than human: The promise and perils of enhancing ourselves*. New York, NY: Oxford University Press.

Cartwright, Nancy. 2007. *Hunting causes and using them: Approaches in philosophy and economics*. Cambridge University Press: New York, NY.

Danielson, M.L., R.H. Bitsko, R.M. Ghandour, J.R. Holbrook, M.D. Kogan, and S.J. Blumberg. 2018. Prevalence of parent-reported ADHD diagnosis and associated treatment among U.S. children and adolescents, 2016. *Journal of Clinical Child & Adolescent Psychology* (2): 199–212. https://doi.org/10.1080/15374416.2017.1417860.

Dees, Richard H. 2007. Better brains, better selves? The ethics of neuroenhancements. *Kennedy Institute of Ethics Journal* 17 (4): 371–395.

DellaVigna, Stefano, and Ulrike Malmendier. 2006. Paying not to go to the gym. *American Economic Review* 96 (3): 694–719. https://doi.org/10.1257/aer.96.3.694.

Douglas, T. (2008). Moral enhancement. *Journal of Applied Philosophy* 25: 228–245. https://doi.org/10.1111/j.1468-5930.2008.00412.x.

Erler, Alexandre. 2017. The limits of the treatment-enhancement distinction as a guide to publicpolicy. *Bioethics* 31 (8): 608–615.

Ferrero, Luca, ed. 2022. *The routledge handbook of philosophy of agency*. Routledge: New York, NY.

Gross, J.J. 2015. Emotion regulation: Current status and future prospects. *Psychological Inquiry* 26 (1): 1–26. https://doi.org/10.1080/1047840X.2014.940781.

Holton, Richard. 2009. *Willing, wanting*. Waiting: Oxford University Press UK.

Ilieva, I., and M. Farah. 2013. Enhancement stimulants: Perceived motivational and cognitive advantages. *Frontiers in Neuroscience* 7 (198): 1–6. https://doi.org/10.3389/fnins.2013.00198.

Kjærsgaard, Torben. 2015. Enhancing motivation by use of prescription stimulants: The ethics of motivation enhancement. *AJOB Neuroscience* 6 (1): 4–10. https://doi.org/10.1080/21507740.2014.990543.

Levy, N., T. Douglas, G. Kahane, S. Terbeck, P.J. Cowen, M. Hewstone, and J. Savulescu. (2014). Are you morally modified?: The moral effects of widely used pharmaceuticals. *Philosophy, Psychiatry, & Psychology* 21 (2): 111–125. https://doi.org/10.1353/ppp.2014.0023. PMID: 25892904; PMCID: PMC4398979.

Mauss, I.B., S.A. Bunge, and J.J. Gross. 2007. Automatic emotion regulation. *Social and Personality Psychology Compass* 1: 146–167. https://doi.org/10.1111/j.1751-9004.2007.00005.x.

Pickard, Hanna. 2011. Responsibility without blame: Empathy and the effective treatment of personality disorder. *Philosophy, Psychiatry, Psychology* 18: 209–224.

Ritchie, S. J., and E. M. Tucker-Drob. 2018. How much does education improve intelligence? A meta-analysis. *Psychological Science* 29 (8): 1358–1369. https://doi.org/10.1177/0956797618774253. Epub 2018 Jun 18. PMID: 29911926; PMCID: PMC6088505.

Sandel, M. 2009. *The case against perfection: Ethics in the age of genetic engineering*, 1st edn. Belknap Press: An Imprint of Harvard University Press (September 30, 2009).

Schüll, Natasha D. 2012. *Addiction by design: Machine gambling in Las Vegas*. Princeton, NJ: PrincetonUniversity Press.

Shepherd, Joshua. 2017. Halfhearted action and control. *Ergo: An Open Access Journal of Philosophy* 4.

Trahan, L. H., K. K. Stuebing, J. M. Fletcher, and M. Hiscock. 2014. The Flynn effect: A meta-analysis. *Psychological Bulletin* 140 (5): 1332–1360. https://doi.org/10.1037/a0037173.

Wiseman, Harris. 2016. *The myth of the moral brain: The limits of moral enhancement*. MIT University Press: Cambride, MA.

Zarpentine, Chris. 2013. 'The thorny and arduous path of moral progress': Moral psychology and moral enhancement. *Neuroethics* 6 (1): 141–153.

Steven McFarlane is an independent scholar who has taught at Florida State University, University of Central Florida, and the University of Minnesota, Morris. His work spans across issues in philosophy and psychology, including questions related to rationality and intentional action.

Chapter 9
Embodiment Diffracted: Queering and Cripping Morphological Freedom

Joshua Earle

Abstract Transhumanists offer as a human right, the duty to allow all people to alter their bodyminds (or not) as they see fit (so long as it doesn't hurt anyone else) without suffering any legal or socioeconomic repercussions. On the face of it, this seems like an incredibly positive and egalitarian human right that has the potential to stop discrimination against marginalized people. However, transhumanists also insist on treating everyone as an individual, and not as a member of any "arbitrary" demographic or group. By disregarding demography, and insisting on treating everyone as an individual, transhumanists fail to appreciate the power of shared embodiment, both politically, socially, and individually. Morphological Freedom thus becomes implicitly focused more on providing rich folks freedom from social mores and laws that may limit their access to technological advantages over non-enhanced people in the marketplace, than on ensuring morphological freedom for everyone. In this chapter, I reaffirm the importance of shared demography as a vital focal point of political power and individual and social health. I discuss how shared embodiment and the power of communal belonging makes an individualistic morphological freedom impossible. Through explorations of communities of disabled people, transgender people, biohackers, and body modders, I re-form morphological freedom to be an expression of communal engagement and belonging, rather than an individualistic and neoliberal competition. The communities I will be discussing lie on a spectrum from individualistic to communitarian, but all of them queer (and/or crip) the concept of morphological freedom that transhumanists have put forth. By diffracting the right of morphological freedom through communities which resist morphological norms, I illuminate both the political limitations of an individualistic morphological freedom, and the political and communitarian possibilities of kinship and belonging within and across demographic lines.

This work is a portion of my Dissertation project, titled: The Myth of Multiplicity: Morphological Freedom and the Construction of Bodymind Malleability from Eugenics to Transhumanism.

J. Earle (✉)
Virginia Polytechnic Institute and State University (Virginia Tech), Blacksburg, USA
e-mail: jearle@vt.edu

9.1 Introduction

Sentient entities agree to uphold morphological freedom—the right to do with one's physical attributes or intelligence whatever one wants so long as it does not harm others.

...[1]

Morphological freedom entails the duty to treat all sapients as individuals instead of categorizing them into arbitrary subgroups or demographics, including as yet undefined subcategorizations that may arise as sapience evolves.

However, the proper exercise of morphological freedom must also ensure that any improvement of the self should not result in involuntary harms inflicted upon others. Furthermore, any sentient entity is also recognized to have the freedom not to modify itself without being subject to negative political repercussions, which include but are not limited to legal and/or socio-economic repercussions.

~Transhumanist Party Bill of Rights, Ver. 3.0, Article X

Morphological Freedom—the "right to do with one's physical attributes or intelligence whatever one wants so long as it does not harm others" (Transhumanist Party Bill of Rights—Version 3.0, Article X)—is central to the Transhumanist movement. Drawn from the logical extension of the Lockean rights of life, liberty, and property (More and Vita-More 2013; Earle 2021a), transhumanists see the ownership of one's own body as absolute, and as such, allow for the alteration of that body, however one might want; just as one might be free to modify one's house to add a porch, or put a spoiler on one's Honda Civic. On the face of it, this seems like an incredibly positive and egalitarian human right that has the potential to stop discrimination against marginalized people, particularly those who have non-normative bodyminds,[2] be they racialized, disabled, trans, etc. This is certainly how transhumanists view the right (Anders Samberg, in More and Vita-More 2013; Fuller 2016). However, transhumanists also insist on treating everyone as an individual, and not as a member of any "arbitrary" demographic or group. By disregarding demography, and insisting on treating everyone as an individual, transhumanists fail to appreciate the power of shared embodiment politically, socially, and individually. Similarly, through aims such as eliminating disability, transhumanists fail to value the morphological variety that already exists, putting lie to any claim that this right could increase that variety. Morphological Freedom thus becomes implicitly focused more on providing rich folks freedom from social mores and laws that may limit their access to technological advantages over non-enhanced people in the marketplace, than on ensuring morphological freedom for everyone.

---

[1] I have omitted the second clause of this right to avoid confusion as it pertains to cryopreservation, a topic not discussed in this chapter.

[2] I use the term "bodymind" in this chapter in the tradition of Black Feminist and Disability Studies (see: Schalk 2018; Price 2015). "Bodymind" subverts the usual separation of body from mind in the Cartesian tradition, and further troubles the association of minds with rationality and masculinity, and of bodies with animality, emotionality, femininity, and racialization. "Bodymind" foregrounds how both are inextricably entangled into a whole that cannot be separated.

In this chapter, I affirm the importance of shared demography and care as vital focal points of political power and individual and social health. I discuss how shared embodiment and the power of communal belonging makes an individualistic morphological freedom impossible. Through explorations of communities of biohackers, body modders, trans people, and disabled people (respectively in the sections that follow) I re-form morphological freedom to be an expression of communal engagement and belonging, rather than an individualistic and neoliberal[3] competition. I then argue that by embracing an ethic of care and my own theory of radical intra-dependence, and deconstructing or abolishing the very notion of the neoliberal individual (and the capitalism upon which it rests), we might actually be able to reach a version of morphological freedom that allows us to thrive in all of our demographic variety.

The communities I will be discussing *queer*[4] and/or *crip*[5] the concept of morphological freedom that transhumanists have put forth. By diffracting the right of morphological freedom through communities which resist morphological norms, I illuminate both the political limitations of an individualistic morphological freedom, and the political and communitarian possibilities of kinship and belonging within and across demographic lines.

I must start here with a caveat. I do not reside in any of the bodymind arrangements about which I write in this chapter. That brings with it some dangers, about which I am particularly wary. The first is that, as someone whose bodymind is among the most privileged possible (cis-het white male), I run the risk of being paternalistic toward these communities by speaking *for* them (or seeming to), rather than raising their voices about their own lived experience. The second is that, in a chapter that discusses four different communities, each of which is nowhere near homogenous, there is literature I am sure to have missed. This work is intentionally interdisciplinary, so I run the very real risk of leaving out important work that may have been covered in any of the fields into which I can only dip a toe for this piece. This is especially true of newer work by marginalized people (ostensibly the people about which I write), which is often overlooked in the academy in favor of the (admittedly important and context-building, but mostly white, male, cis, and abled) "official canon." I risk both

---

[3] "Neoliberal" in this context describes the combination of the "Classical" liberalism of philosophers like John Locke, René Descartes, Emmanuel Kant, and others with the capitalist construction of the individual as a measure of labor value that can be extracted. For a more nuanced definition and history of the term, see: Thorsen (2010).

[4] The moniker "queer" in this case is a reclamation of the slur into an adjective/verb in queer theory that means "to use against the dominant or normative narrative, or to use in a way that subverts expected narratives."

[5] Similar to "queer," to "crip" something is to approach an issue from the positionality of a disabled bodymind. This also reclaims a slur once used to demean, but is now used to expand the possibilities the thing being cripped might afford. Crip time, for instance is used to describe and disrupt "normal" expectations of productivity, punctuality, and endurance to include those who may only have energy for short periods of activity, may be interrupted by flare-ups of disease, or who might be delayed by faulty equipment, inconsiderate parkers, or other realities of the disabled lived experience. See: Samuels (2017).

reinventing a wheel already firmly established by someone else, to building wheels that no one in these communities asked for.

9.2   Biohackers

Biohackers are a group of people (mostly, but not entirely, white and male) who grew out of the cyberpunk movement of the 1980s and 90s. They are perhaps the single group most aligned with transhumanist ideology, and many if not most likely align themselves with (or identify directly as) transhumanists. Biohackers are also part of the DIY/open science movement, and use publicly-available tools to do genetic science (Delfanti 2013). Some of the more extreme practitioners attempt to produce genetic therapies (to do things like increase muscle mass or cure/prevent Herpes), and experiment with them on themselves. They also, following their cyberpunk roots, like to implant technological devices under their skin that do things like activate smart home devices or unlock remote-keyed automobiles. By rejecting the everyday belief that biology is set, they queer the very notion of biological determinism (and, somewhat ironically, the "born this way" political stance that gave the LGBTQIA movement significant purchase in the 2000s).

Biohackers have a lot in common with the "gender hackers" that Preciado claims kinship to (see Sect. 9.4), though few would admit such allegiance. They more often claim kinship with scientists like Barry Marshall, the Australian scientist who "proved" his hypothesis that stomach ulcers were caused by bacteria (rather than by stress or spicy foods) by ingesting a broth infused with H. pylori to trigger ulcers within himself (Hellstrom 2006). Marshall won a Nobel prize for this work, and such accolades are no small part of why many biohackers do what they do. Biohacking borrows its name from computer hacking, which is the practice of using computers in ways that work against government- or corporate-established rules and norms, often towards liberatory purposes. Open-access to knowledge, and the techniques to produce that knowledge, are the driving ethos' of both computer hackers and biohackers.

Perhaps the most comprehensive piece on biohackers is the book *Biohackers: The Politics of Open Science* by Alessandro Delfanti (2013). Delfanti illustrates the heterogenous ethics and norms of the open science community. Delfanti notes that there are some generalized ethoses to which biohackers adhere: openness (in access to knowledge, materials, and techniques), resistance to codified and siloed academic or corporate certification, and freedom "...as in free speech, not as in free beer," (p. 121). This last is particularly notable, since it is often believed that hackers want all things to be monetarily free, when it is only the knowledge itself that they wish to be freely available (and, free to pursue, hence "speech-not-beer"). In fact, Delfanti highlights the entrepreneurial drives of many biohackers and open science advocates, noting that many seek to use free knowledge to create proprietary things. He outlines a subsection of biohackers specifically as profiteers. Some biohackers find profiteers to be problematic or even antithetical to the project of open science, but many in hacker communities have gone on to become incredibly rich through the projects they produced, Bill Gates being perhaps the most (in)famous example.

How the biohacker community reacted to the meteoric rise and untimely death of Aaron Traywick is particularly telling in how they conceive of their own community and work. Traywick was a young biohacker with a talent for holding extravagant public experiments, most infamously injecting himself, live on stage, with what he claimed was a cure for herpes.[6]

Traywick was a divisive figure within the biohacking community. Many saw his often-rushed attempts at public spectacle as damaging to the image of biohacking. His public antics garnered a rebuke from the FDA, which warned biohackers not to sell any therapies they might have produced (Brown 2017), and far more public media than most biohackers were comfortable with. One major complaint that the biohacking community had was that Traywick made unrealistic promises, on time-lines that were impossible to keep. His (very public) failures threw into question the expertise and validity of open-source biomedical research, and threatened to get the entire operation shut down by Federal regulators. Less than three months after his herpes injection, he was found dead in a sensory deprivation tank, having drowned.[7] He was 28 (Brown 2018).

The particularly libertarian, individual-freedom-meets-cyberpunk-optimism bent of the biohacking community makes this a particularly sticky situation. They see themselves as experts outside the standard reckoning of scientific institutions, and better for it: not shackled by rules established by the FDA, which arguably contribute to the vast majority of drugs failing to reach the market. They see themselves as being able to take bigger risks, and possibly make more spectacular gains because of it. By sticking to self-experimentation, they adhere very much to the transhumanist value of altering oneself without harming others, and also sidestepping FDA regulations, which target medical experimentation on other people.

And yet, without a mechanism to prove that they are, indeed, experts, they remain vulnerable to a "used car salesman" who can paint them all as charlatans at best, and dangerous at worst. While biohackers are unlikely to alter their own germ cells (necessary for producing a heritable alteration), any successful gene alteration carries substantial personal risk, and may also have knock-on effects that could be entirely unforeseen. Traywick's death became a source of a lot of conspiracy theories. Many stemmed from a desire to make sure that biohacking itself wasn't claimed as the cause of his death, and thus regulated away (Brown 2018).

Peter Thiel, founder of PayPal, chairman of Palantir (a software company that deals in big data analytics, and is entwined with government counter-terrorism and widespread surveillance projects), and advisor to former President Trump, is a proponent of biohacking and a hero of the transhumanist community. He is a vocal supporter of using blood transfusions from young donors as a method of staying

---

[6] It is unclear which Herpes complex was the target of his cure.

[7] The exact cause of death beyond drowning is unknown, but his autopsy found ketamine in his system. There is no evidence that any of his self-experimentation directly led to his death.

young (Bercovici 2016),[8] and is actively funding longevity projects. He also wants to fund laboratories which can work outside of the normal bioethics' standards, via a mechanism called "seasteading." By planting a lab in international waters, any research undertaken there would not be subject to a nation's laws or regulations.[9] Thiel's ambitions could scale up biohacking out of basements and into a lab funded by one of the richest men in the world. Issues of expertise, law, governance, personal and national sovereignty, and more could be at stake. How that shakes out, should Thiel attempt to put his seasteading dream into practice, could determine a lot about how biohacking continues, and how transhumanism evolves in the coming decades.

The fundamental ways that successful do it yourself genetic engineering could open up possibilities for alternative bodymind arrangements, and significant individual and social risk make biohackers an urgent community for Science and Technology Studies (STS) and bioethicists to study. The transhumanist ethic of morphological freedom, as it is written currently, with all of the pitfalls I've written about elsewhere (Earle 2021a) is one that aligns almost perfectly with biohacking… and yet the imbrication with people like Peter Thiel only magnify the risks. Should the movement gain some more widespread acceptance, there is the possibility of actual generative multiplicity. But the dangers of biohacking, much like the hacking of personal photos or financial information, and especially when we are increasingly conjoined with our technology, have the potential to meet and multiply. As Elon Musk introduces a neural chip that he claims can help with memory retention, and to cure all sorts of brain diseases[10] (Metz 2019), the bio-technical reality of our cyborg bodyminds means that these issues can no longer be considered separate.

If biohackers are successful in their endeavors, they will inevitably succeed in introducing new morphological possibilities, but they will simultaneously reduce or eliminate others. If Musk's chip can eliminate something like autism (it can't, but for the sake of argument), that is a whole demographic that is now in danger. As much as biohackers imagine themselves the vanguard of new, revolutionary medicine, it seems more that they are looking for a mechanism of getting ahead, and getting around the rules designed to keep people safe. The focus on individual achievement over community thriving, of self-advancement over care, leaves biohackers as both the most aligned with the Transhumanist ideal, and the least likely to produce any morphological freedom that supports a thriving and diverse population.

---

[8] It is not clear if he has actually undertaken this procedure.

[9] Any attempt to market a drug within the US (or other country) would require the research to have been done in ways that upheld national ethics requirements, though, so experimental basic science might be able to be done in a seastead, but development and clinical trials would still need to be done under the auspices of the FDA or other national ethical standards.

[10] It won't. And some of the "diseases" he claims that the chip will cure (such as autism) are not even diseases. Disability scholar Liz Jackson calls this a "disability dongle," a "well-intended, elegant, yet useless solution to a problem [disabled people] never knew we had" (Jackson 2019). Common examples include stair-climbing wheelchairs and gloves which turn sign language into speech.

9.3   Body Modders

One group who is taking a more established (in the sense of using techniques available to people for longer than direct genetic manipulation has been) approach to altering their own bodyminds are body modders. Beginning in earnest in the US in the 1990s—and falling out of the broadening tattoo popularity, especially that inspired by African and Indigenous (Māori especially) practices such as the now-ubiquitous "tribal" tattoos (DeMello and Rubin 2000)[11]—body modding uses techniques gleaned from tattoo artistry, Western cosmetic surgery, African and Indigenous modification practices such as gauged piercings and scarification, to new techniques invented by the movement grinders themselves (Pitts-Taylor 2003). Modders nowadays can have quite unique appearances, from the full-body tattooing of someone like Rick Genest (aka Zombie Boy, made famous by his appearance in the music video to Lady Gaga's song Born This Way) and even scleral tattooing (altering the color of the whites of one's eyes), to tattoos and scars which imitate dragon scales, tooth-filing to produce animalistic fangs, to much more serious procedures such as dermal implants, various bifurcations (tongues and genitalia are the most common), heavy piercing, and even amputations. Body modders, like biohackers, have roots also within the cyberpunk movement, but generally trend more punk than cyber, firmly established within the lower-class anti-establishment ethos of punk. They queer the very notion of morphology, altering themselves in ways far outside the norm. They also crip ideas of morphological normativity both because many already claim disabled identities such as neurodivergence, but also because the modifications themselves can disable, both in the "medical model"[12] definition of altering the everyday functions of their bodyminds, but also in the "social"[13] model of disability in the sense of imposing onto them the kinds of stigma that regularly affect disabled people.

While the techniques of body modification can resemble that of cosmetic surgery, most body modification is not done by medical professionals, nor do they modify bodies towards a single vision of human beauty as cosmetic surgeons do. It is nearly impossible to find a reconstructive surgeon, cosmetic surgeon, or other medical

---

[11] That a significant portion of the modification community stems from what Juno and Vale (1989) call "modern primitivism" tied to an image of tribalism and indigeneity, while also being, as Pitts remarks, "a white, gay-friendly, middle-class, new-age, pro-sex, educated, and politically articulate set of people" (2003, pp. 13–14) (Musam Fakir, often credited with starting the modern primitivism movement, was white as well), is an interesting dichotomy. How problematically appropriative modders might be - themselves quite stigmatized, though often from privileged backgrounds—is a discussion beyond the scope of this chapter. Pitts discusses this tension in Chap. 4 of her book.

[12] The "medical model" of disability claims that the site of disability is some abnormality or injury to a person's bodymind, and that the solution to said disability is a medical alteration or "fix" towards normativity. For example, someone with a spinal injury would be disabled because of that injury and the solution would be the use of a wheelchair or other mobility device (see: Adams et al. 2015).

[13] The "social model" of disability argues that bodyminds are not the sight of disability, but rather the architectural and attitudinal barriers that society puts up in order to keep disabled people out of regular public life. To continue the wheelchair example, the social model argues that it is not the injured spine that causes the wheelchair user's disability, but rather the ubiquity of stairs and stigma that marks the wheelchair user as "other."

professional who is willing to alter a bodymind significantly outside of what is considered "normal." Since modding is usually done in order to separate one's appearance from the norm, those who do it must find alternatives to established medical "experts." Most body modification is done by people known as "grinders." Often self-taught or taught through apprenticeship within the body mod community. Grinders, while undeniably skilled, usually have little to no formal medical training.

Victoria Pitts-Taylor's first book, *In the Flesh: The Cultural Politics of Body Modification* (2003),[14] follows the rise of body modding alongside the cyberpunk boom of the 1980s and 90s. In the more than decade and a half since it was written, it has perhaps become slightly out of date for describing social structures within the movement, but can give us some insight into where the movement came from, and many of the struggles still associated with it. Pitts describes body modification as:

> [emerging from] a network of overlapping subcultural groups with diverse interests [including kink and fetish, sex-positive feminists, punk, and queer groups] who eventually began identifying themselves and each other as "marked persons" or as body modifiers. What they shared was that they all positioned the body as a site of exploration as well as a space needing to be reclaimed from culture… The affective aspects of the body—for instance, its experiences of pain and pleasure in sexual practice and in non-Western tribal rituals—and its political significance became a primary focus of body modifiers. Instead of an object of social control by patriarchy, medicine, or religion, the body should be seen, they argued, as a space for exploring identity, experiencing pleasure, and establishing bonds to others (pp. 7–8).

The focus on sex, and the use of modification as community binding here are significant. Similar to how queer, transgender (Lawrence 2004; Worthen 2016), and some disability communities (Elliott 2003; Gilbert 2003) often have some form of deviant sexual pleasure associated with them—which is stigmatized by (particularly American) puritanical heteropatriarchal systems—the modification community embraces both the sex and the deviance within such denigrations. They reclaim the stigma as a core identity—valorizing the crossing of affective boundaries, confounding pleasure and pain—and produce a community which fundamentally resists categorization or Western neoliberal standards of being.

While transhumanists often consider morphological freedom a release on the limits of how we might alter the body, and thus find out what we might be able to make the body *do*, modders are generally more interested in what we can make the body *feel*, *be*, and *mean*. The things that transhumanists avoid when including modders within their concepts of morphological freedom are twofold. First, transhumanists tend to eschew[15] any discussion of the affective (i.e. feeling) dimensions of body modification. The philosophical foundations of transhumanism have a fundamentally masculinist bent, and most transhumanists deride, if not outright reject, the notion of feeling or emotion (i.e. feminization,) as important in any way. Transhumanists tend to align entirely with an Enlightenment notion of reason as the ur-quality which

---

[14] Published under the name Victoria Pitts.

[15] Gesundheit.

makes humans human.[16] Julian Savulescu, a transhumanist bioethics professor, wrote in one article that "…we are entering a new phase of human evolution—evolution under reason[17]—where human beings are masters of their destiny" (2003). Zoltan Istvan explicitly removed any reference to emotion when appropriating the right of Morphological Freedom for the Transhumanist Party, and has vocally denigrated emotion as frivolous (Earle 2021a). The disconnect from the primarily affective nature of body modification indicates that transhumanists see modders simply as convenient visual props for their pro-technological arguments without including the ways that modders produce and uphold their form of affective community.

Second, transhumanists also avoid discussing how much social stigma and exclusion body modders face in broader society. Such exclusion runs rather counter to their argument that modders are evidence that society will accept a wide variety of morphological assemblages.[18] Pitts-Taylor discusses early in her introduction Andrew, a young body modder who, due to his "shocking" appearance, cannot find employment outside of being a self-employed tattoo and piercing artist, avoids seeking medical care for fear of being labeled mentally ill and forcibly institutionalized, and was also denied admittance to Canada (2003, p. 2). She also notes that modders often find themselves employed as tattoo and piercing artists, or in places where "their modifications enhance, rather than eliminate, their prospects" (2003, p. 12). While prominent visible tattoos on the face and hands are becoming more popular among celebrities such as rapper Tekashi69 and pop artist Post Malone, tattoos visible outside of a "business-casual" dress style (slacks, long-sleeve button-up shirts) will generally prevent one from being able to get many jobs outside of retail and service industries. The more severe modifications making even those positions unlikely.

Medical care for modders and other non-normative-looking people is incredibly difficult. Modders have similar issues to trans people (and, frankly Black and brown people as well, though the modder community is majority white) in that their appearance can often catch doctors unaware, and often trigger feelings of disgust and other aversions that can negatively affect modders' medical care, or even cause doctors and emergency services to withhold care altogether. There are many news stories of trans women who were denied care by emergency workers and died from that neglect, and if Pitts' description of Andrew is any indication, it is likely not unknown within the modder community as well. Due to their decisions to implant technologies, and alter their bodyminds away from the "normal" in medical contexts, they can find their concerns either not believed—due to some believed mental deficiency that would be the cause of the desire to mod—or have their medical issues considered to be their own fault, and thus deserved, and not worth treating. There is rhetoric about the

---

[16] See: Bacon, "Knowledge itself is power" (1597); Kant, *The Critique of Pure Reason* (1781); Paine, *The Age of Reason* (1794, 1796, 1811), and more.

[17] This phrase, "evolution under reason" echoes a tagline for the eugenics movement in the early twentieth century: "the self-direction of human evolution." It appears most famously on the poster for the 1924 Eugenics Conference, an image most easily found on the Wikipedia page for eugenics.

[18] Anders Samberg does note that it is highly likely that social custom will remain clustered around certain accepted norms under Morphological Freedom, but maintains that variations, even if rare, would not be stigmatized (More and Vita-More 2013, p. 59).

modder community of "self-mutilation" from medical circles and society in general that works to denigrate not just modders' bodymind choices, but serves to call into question their mental health as well. Add to that the modern tribalism, and the echoes between "self-mutilation" and the practice of Female Genital "Mutilation" or "Cutting,"[19] and the modding community gets further stigmatized and racialized (even though the majority are white).

To counteract a lot of this stigma, the body modder community is, internally, incredibly strong, even if relatively small. Most modder communities congregate in large metropolitan areas where it is both easier to find like-minded others (including those with the skills to perform the modifications), and also to have a community of the size necessary to support each other. The mutual aid and community care that they provide for each other is remarkable. They collaborate on medical knowledge; maintaining their implants, piercings, and other mods.[20] They establish and maintain employment networks for each other, and also maintain mutual aid networks, and even communal living arrangements. These communities hold together through both a shared ethos of self-expression and counter-culture transgression of norms, but also through support, care, and an affective connection to those who share their beliefs. These affective connections, the power of shared demographic, especially as a demographic against normative power structures becomes the glue holding these communities together.

9.4 Trans Communities

As one of the few communities I discuss whose bodymind alterations are currently done by licensed medical professionals,[21] transgender people's experience with the medical system is incredibly important in this discussion. Similarly, trans communities' experience with gradually growing social acceptance—juxtaposed with the physical violence they face from those most invested in maintaining a normative status quo—puts them at the center of many of the right-wing bad faith "culture wars" debates as the edge case which will lead down the slippery slope to such things as bestiality and pedophilia. That trans women of color are murdered at rates almost unheard of in other demographics (Forestiere 2020) makes clear that how society deals with non-normative bodymind arrangements, and especially those

---

[19] Activists around FGM or FGC prefer the term "cutting" rather than "mutilation" because of the negative moral valence of the term mutilation. That association also ends up applied to body modders.

[20] This is a practice, along with how disabled people engage in the medical industrial complex, that I call cyborg maintenance. Cyborg maintenance foregrounds how the integration of technology and biobodyminds both solves and creates problems. Maintaining these intersections produces new relationalities, and new vectors for care and neglect.

[21] This has not always been the case, of course. In many ways, trans communities could be considered the first body modders, as they often had to seek out black-market options for altering their bodyminds.

who choose to alter themselves against those normativities, is of utmost importance in producing an ethical and just morphological freedom. Trans communities most obviously associate with the queering of morphology, and of morphological stability.

The limited but growing social acceptance of transgender communities is behind the move by the Body Integrity Identity Disorder (BIID) community to claim the "identity disorder" language of the trans community, as well as often calling themselves "trans-abled."[22] Both the Trans community, as well as the BIID community, made these rhetorical moves because of a widespread belief, and official diagnoses via the medical system, that they were both simply paraphilias. Homosexuality, transgenderism, and BIID were all originally classified as a pathological sexual fetish: a problem of abnormal sexual fixation. The medical establishment believed that those with the conditions of transgenderism or BIID could only (or mainly) find sexual satisfaction through the "aberrant" behavior, be that sex with one's own gender, altering one's sex, or becoming disabled respectively (Elliott 2004; Gilbert 2003).

And yet the widespread backlash and violence against trans people (particularly trans women of color) illustrates that medical diagnosis and identity language are not enough to produce social acceptance. Technoscientific classification and objective data have never been sufficient cause for society to grant personhood or equality to a particular group (see: the constant attempt of genetic scientists to eradicate the notion that race is genetic and determinate of behavior, vs the evolutionary psychology field as a whole). This does not bode particularly well for the BIID community, but in more general terms, it also strikes a considerable blow to the transhumanist notion that the technical ability to produce a wide variety of bodymind arrangements will necessarily lead to the acceptance of a multitude of those arrangements.

But more important than any destabilization of transhumanist assumptions is the lived experience of being trans which opens up avenues of multiplicity and bodymind malleability, the kinds of communities such malleability can produce, and the complex and contested relationships with sociotechnical systems that they reveal. To this end, I turn to two trans scholars, Eli Clare and Paul Preciado,[23] both of whom have written extensively on their own transitions and the kinds of struggles and communities that they encountered. I turn to these authors not only because of their insight and relevance to my own arguments, but because they both connect directly to the other communities I discuss here: Clare to disability communities, Preciado to biohackers and body modders. They are emblematic of how none of these categories (queer, disabled, gendered, etc.) is absolute. We all embody and inhabit various positionalities, many of which intersect in ways that have profound effects on our engagement with the world. To be disabled and trans is a fundamentally different thing than being disabled and cis, or trans and abled. It is not merely the

---

[22] I have reservations about using the term "trans" here, but discussion of such is beyond the scope of this chapter.

[23] That both of these scholars are white and trans-masculine is of some concern to me, as it is usually easier to "pass" and be accepted when presenting as masculine. Thus, neither author has had the particularly precarious experience of being a trans woman, nor as a trans person of color. This limits how generalizable their accounts can be considered, even beyond the particular time/place in which they experienced their various gender identities and/or transition.

sum of its parts, but a new ontology altogether (cf. Crenshaw 1989, 1991; Collins 2000, etc.).

Something both authors teach us is that transition is a process, forever in progress, never complete. It is an embodied experience that puts the lie to any notion of a binary within sex, gender, or really any clean division between identities. Clare describes his transition as:

> ...a long meandering slide. Today I live in the world as a man, even while my internal sense of gender is as a genderqueer, neither man nor woman. At the same time, I have no desire to abandon or disown my long history as a girl, a tomboy, a dyke, a woman, a butch (2015, xxvii).

Clare, though he does not shy away from his queerness or his transness, does not focus on it in his writing. It becomes apparent that he never ran up against his gender in the same way that he did his disability. He writes about wanting to cut off his right arm to stop it from trembling (p. 151), and yet his transition from girl to man is described as a long slide, apparently without much in the way of incident or bump. The way that disability becomes foremost in his writing indicates that his gendered experience is of a kind that some others cannot enjoy: one of easy movement between socially-legible locations.

Another way that Clare's experience of Transness troubles the "common sense" understanding of gender itself,[24] is that the fluid motion between gender expressions, and his insistence that he was once a girl, a butch, and a dyke, foregrounds the possibility of change in personal identity, not just of one's bodymind materiality to match one's always-already static identity. Even in LGBTQ+ communities, the "born this way" mantra is often used as a political tool to resist the stigmatized critique that homosexuality or queerness is a choice, and thus the fault of the person who chooses to take on a socially-taboo identity for any stigma they endure.[25] Often, the stories we hear of trans people are ones of discovering the person they have always been, having been born in the wrong body, or some other version of having a static gender that does not line up with their birth-assigned gender. Clare's story shows us

---

[24] The academic and activist positions on the fluidity of gender is not settled. Parlance in public, in news stories, or (god forbid) online tends to maintain that identity categories such as gender and race are fixed, even if they don't map on to biosex. This is somewhat true even in Trans communities, as evidenced by the controversy around detransitioning I mention in footnote 22.

[25] The rise of CRISPR and the possibility of editing embryos to remove undesirable traits (see: Isaacson 2021) has destabilized the notion of a static, "born this way," identity even further. Combined with the many studies trying to find the genetic causes of homosexuality (see: Turner 1995; Rice et al. 1999; Reardon 2015; and many more), this destabilization opens up new avenues of the democide (see: Rothblatt 2011) of LGBTQ+ people. It also, if we want it to, could lead to MORE avenues for morphological freedom and variation. If widespread genetic editing of embryos becomes possible without any associated social justice movement for LGBTQ+ people, it will most likely lead to more homogeneity as people (understandably) work to keep their children out of the pain that comes from being subject to homophobia, transmisia, and other oppressions. For these reasons, LGBTQ+ rights activists have all but abandoned the "born this way" mantra.

that this need not be the case. One's gender identity and/or expression need not be fixed for one's entire life.[26]

Preciado describes his transition in a much more disjointed way than Clare. He describes swinging wildly back and forth at the whims of "T" (testogel or topical testosterone) thus:

> [A]n extraordinary lucidity settles in, gradually, accompanied by an explosion of the desire to fuck, walk, go out everywhere in the city... Absolutely all the unpleasant sensations disappear. Unlike speed, the movement going on inside has nothing to do with agitation, noise. It's simply the feeling of being in perfect harmony with the rhythm of the city. Unlike with coke, there is no distortion in the perception of self, no logorrhea or any feeling of superiority. Nothing but the feeling of strength reflecting the increased capacity of my muscles, my brain. My body is present to itself. Unlike with speed and coke, there is no immediate comedown. A few days go by, and the movement inside calms, but the feeling of strength, like a pyramid revealed by a sandstorm, remains (2013, p. 21).

Preciado frequently depicts these episodes of near mania once a regimen of testogel is administered, describing desires that we politically ascribe to masculinity; wanderlust, physical lust, the desire to fuck and possess. He maintains that testosterone isn't masculinity itself (p. 141), but also that *clinical* masculinity cannot exist without synthetic testosterone (p. 61). Synthetic testosterone opens up political possibilities of masculinity, once only afforded to cis-men.

The kind of black-market sharing of T that Preciado discusses shares a striking resemblance to how AIDS activists in the 90s shared HIV drugs, and were forced to gain and assert their own expertise in the science of HIV medication (Epstein 1995). Preciado describes the illicit way in which he obtained testogel from black market dealers, describing his own addiction to testogel, all signifiers of drugs such as cocaine, amphetamines, or heroin. The experimentation he describes, he also compares to the way people experimented with hallucinogenic as mechanisms of healing, spirituality, and witchcraft (Preciado and Benderson 2013, pp. 145–151). Experimentation necessarily falls outside of the biomedical regimes in which one might find themselves, as well as the capitalist structures that hold those biomedical regimes up. Instead, a much looser, trade-based, often on credit, often with the promise of future favors, and often with money gleaned from other subversive and/or illegal actions such as sex work or other drug trafficking, mechanism of capital is employed.

---

[26] There is also the experience of people who de-transition. While sometimes this is done to avoid stigma and violence (particularly if it is difficult for the person to "pass" as their desired gender, other times it is because they have simply changed again. De-transitioners are controversial in trans communities, because they are easy ammunition by TERFs (Trans-Exclusionary Radical Feminists, an anti-trans academic movement who use biological reductionism and a direct mapping of biosex to gender to deny trans people rights) and other bad-faith groups who want to argue that transness is just a phase, or some aberrant behavior that needs to be curtailed. Trans folks can often feel betrayed by a member of their community who decides to "switch back," and the cognitive dissonance is one that has yet to be fully theorized in the literature. In fact, searches for literature on detransitioning bring up mostly anti-trans screeds and discussions of families "detransitioning" trans family members to their deadnames for funerals and obituaries after they have died.

These mechanisms of subverting the capitalist distribution of care is also a factor in what Preciado claims is the goal of the pharmacopornographic era. He writes: "Pharmacopornographic capitalism is ushering in a new era in which the most interesting kind of commerce is the production of the species as species, the production of its mind and its body, its desires and its affects." (2013, p. 51). This production is interesting not just because it is creating ways of bodymind being and affect, but also, the notion of producing the species itself, i.e. the category of being that is singular and separable from other modes of being. In a way, the mechanism by which we can shift from one kind of body to another, simultaneously maintains the separability of the human as species, and, perhaps, the gender binary as (separate) species as well.

And, while this may be the case (or *a* case within multitudes), Preciado considers his community to be about breaking out of pharmacopornographic capitalism. He, instead is a part of a community of gender pirates or gender hackers. He writes "I belong to this… group of testosterone users. We're *copyleft*[27] users who consider sex hormones free and open biocodes, whose use shouldn't be regulated by the state or commandeered by pharmaceutical companies" (2013, p. 55, italics and footnote in original). These remixers, who yearn to be free of the biopolitical, technocapitalist, and disciplinary boundaries drawn by medical and scientific regimes share a lot with the biohacker movement.

Preciado and Clare both illustrate how there are a huge number of ways to be trans or queer, that need not even have all that much to do with one's morphology. This, of course, aligns perfectly with the transhumanist belief that we ought to be able to alter our bodyminds (or not) as we see fit, yet trans people are often excluded from transhumanist spaces.

9.5   Disability Communities

Disabled people have been struggling to be accepted as valuable members of American society for decades if not from the time the first colonizers set foot on the continent. As a key target of the eugenics programs of the early twentieth century, and of the Nazi regime, their very survival has often been threatened by the ableism in which all of society stews. Alternatively seen as objects of pity, or moochers on society, as well as inspiration for ableds to understand that their lot in life could always be worse, disabled people are simultaneously vilified for failing to live up to the neoliberal model of the economically-productive citizen, and held up as emblematic of the gritty, can't be held down, bootstrappy paragon of that same neoliberal spirit. Depicted as always already an individual fighting against their own bodyminds, always already striving to achieve "normalcy." By claiming their disabled bodyminds as normal and good and desirable, disabled activists reject this framing, cripping their own social station, and the idea that "fixing" their disabled bodyminds should be their obvious goal.

---

[27] A play on the term copyright.

I include disabled people in this chapter for multiple reasons, even though most do not choose their particular bodymind arrangement. Disabled people—though they are not a monolithic group, and they are not immune from internalized ableism themselves—embody, quite literally, alternative morphologies that ought to be valued by transhumanists according to morphological freedom. Without valuing morphologies which go against norms now, how does a society accept morphologies which break those norms, albeit in new ways, in the future? Instead, transhumanists revel in the idea that their movement wants to end disability altogether (See, Hugh Herr in Brashear 2014; Istvan 2015; Savulescu and Kahane 2011). Disabled scholars and activists actively resist these narratives of cure and fixing (Clare 2017; Puar 2017). I also include Disabled communities here because they are, themselves, biohackers and body-modders.[28] Arguably the very first of both. And their care practices, which involve modding and hacking, can teach us a lot about both what it is actually like to have an alternative morphology, and also what accepting and valuing those morphologies looks like. Through these associations, we begin to see how care and morphological freedom are entangled, and perhaps, even begin to see a way to produce a morphological freedom worth wanting. And finally, I include disabled communities because they also show us just how toxic, and anathematic to morphological freedom, neoliberal subjectivity truly is.

Disability communities are mostly populated by individuals who did not choose to become disabled,[29] unlike the rest of the communities I discuss. The vast majority of disabled people were either born disabled, or became disabled through accident, disease, or old age.[30] In fact, Rosemarie Garland Thompson famously stated that "Disability is an identity category that anyone can enter at any time, and we will all join it if we live long enough" (2002, p. 20). Those who do choose to become disabled, either through significant body modification (see Part 2), or due to direct action by individuals to disable themselves,[31] are often excluded from—or actively held up as pariahs to—Disability communities. Weise (2016) lumps them in with transhumanists (who desire to become cyborg) as "tryborgs," a portmanteau play on the terms "cyborg" and "try-hard." Try-hard is a derogatory term for someone who wishes to become something for which they are fundamentally incapable or unqualified. "Tryborg" seeks to point out false authenticity, to exclude people from

---

[28] Any of these categories can be inhabited by anyone, including multiples, of course. You can be a Disabled, Trans, Queer, Body Modder. In fact, many of these categories often overlap.

[29] The number of people who do so is likely vanishingly small, to be fair. Actual numbers are not even known well enough to give an estimate of the incidence within various populations. The limited knowledge about the condition points to a low incidence, but it is theorized that it works via similar mechanisms as other dysphorias, so may be similar in incidence. Numbers for gender dysphoria vary, but seem to range from as high as 1.2% (Clark et al. 2014) to as low as 0.002% (via. DSM 5).

[30] The conflict between aging and disability is fraught within the literature. Older people generally resist a disability status even if they might qualify, often because of the stigma that disability has.

[31] See: Carl Elliott's *Better Than Well*, Chapter 9: Amputees by Choice (2003), and the documentary Whole (Gilbert 2003).

adopting a politically perilous and fragile identity, and tries to protect the "actual" community of disabled people from pretenders.

Disability communities feel they need to protect themselves in this way because the "moocher" narrative is one which is often used in order to deny them significant assistance from State and Federal entities. This moocher narrative also comes from the kind of rugged individualism at play in transhumanist circles—but also more widely in American culture—that we call "neoliberal subjectivity." Neoliberal subjectivity is the particular way (classical) liberal individualism gets caught up in capitalist production, and how both of those produce a certain expectation of personhood and value. Disabled people generally have difficulty achieving all of the requirements to become a true neoliberal subject. The very definition within the Americans with Disabilities Act (ADA) which says "impairment that substantially limits one or more of the major life activities," (42 U.S.C. § 12101). It is often the case that a real or perceived inability to work a full-time job is itself a "major life activity." Even if a disability may not impair one's capabilities at a job, disabled people face incredible discrimination in employment. Some are unable at all to participate in economic production. Those that are employed can actually be paid subminimum wages (via Section 14c of the Fair Labor Standards Act), and make on average $3.34/h. according to a report by the U.S. Commission on Civil Rights (Selyukh 2020). Monetary assistance for disabled people is also incredibly limited, and insurance, even after the Affordable Care Act forbade the alteration of prices based on pre-existing conditions, can be incredibly expensive. Not to mention the cost of acquiring the necessary technologies and assistances that being disabled can require, such as mobility devices, prosthetics, medications, in-home care attendants, and more. Power wheelchairs can cost more than most mid-range automobiles, and altering one's own automobile to accommodate that same chair can similarly run into the thousands of dollars. When these technologies inevitably break down, repair can be difficult. Prosthetics usually require appointments with a prosthetist, which may require significant travel and time off from one's job, depending on where one lives. Power chairs and heavily-computerized prosthetics can need to be sent away for repair, sometimes for extensive periods of time, during which the disabled person can be stuck at home with no way to leave for their job or other necessities. These manifold and monumental barriers to becoming a full neoliberal citizen produces disability as something to avoid, as something undesirable, and something to be technically (i.e. via neoliberal innovation) "fixed."

Disabled people have always seen these cracks, and these possibilities (Wong 2020). Medical insurance tied to employment is a mechanism of control, forcing people into the workforce. This can result in disabled people (and nondisabled people, to be fair) overextending themselves in order to maintain full-time status just to be able to afford medical care and medication. Disabled people have long bemoaned the sorts of binds—make too much money -> lose disability benefits; get married -> lose disability benefits; lose job -> lose health insurance—which capitalism has put upon them. The rest of us feel those binds as well. How many parents have missed special moments with their children because they "had" to work? How many times have you missed an important event because you either couldn't afford the trip or couldn't take time off work? How many times did you or someone else go to work sick, and then infect several other people, because you didn't have access to sick leave? How many of the people you infected were chronically ill, disabled, immune-compromised, or otherwise more likely to get a severe version of the disease? Disabled people already know all too well the risks associated with neoliberal subjectivity. The COVID-19 pandemic is slowly teaching the rest of us, but we ought to listen to those for whom this has been the reality for much longer. They can teach us a lot. As Haraway (1991) explains, the position of marginality is a marker of expertise.

As I discussed in the previous sections, we already fail to value much of the morphological multiplicity that exists. I mentioned above how disabled people already get stuck in a multi-bind of being kept out of workplaces and other areas—often literally in the case of inaccessible buildings and public areas. They also get held up as the folks who are both brave to be out shopping like folk need to do, and yet also bootstrappy and not letting their disability hold them back. They become commodified as inspirations for the able bodied (see: Stella Young's 2014 TED Talk on this). This ends up requiring a very complex set of performances whenever disabled people find themselves in public, attempting to go about their day (see: Kasnitz 2020). Requiring this performative song and dance from disabled people is not the same as acceptance, and is very, very far away from valuing their existence. And only through valuing disabled experience and disabled bodyminds can morphological freedom ever come to be.

9.6 Care

To disrupt this neoliberal narrative, I'd like to discuss how the communities I describe here illustrate how care can promote the kind of morphological freedom that transhumanists claim to want. Care work is, itself devalued socially and economically; and that devaluation has sympoietically[32] co-produced care work which is racialized, gendered, and classed in a multitude of ways. Nedi Atanasoski and Kalindi Vora discuss the sorts of racialized work in and around care labor in their book *Surrogate Humanity* (2019). Leah Lakshmi Piepzna-Samarasinha also discusses the sorts of marginalizations that care work is given (2018). Aimee Bahng discusses the sorts of racial and class divides that surrogate reproduction is rife with (2018), and Corbett O'Toole discusses instances where access and care break down, and who gets assigned responsibility for those problems (2015).[33]

In-home care work is 88.5% women (racial demographics roughly follow US racial percentages), with certain jobs such as cleaning and laundry services leaning significantly Hispanic and Black (46% of jobs vs. 26% of US population). Nail salons and massage parlors are nearly 50% (east and southeast) Asian women, even

---

[32] "Becoming-with" via Haraway (2016).

[33] Care is also gaining popularity in other areas of scholarship as well. See: Wittkower (2016, 2020) in Internet Studies, and Herdegen (2019a, b) in Maintenance Studies.

though people of Asian descent only make up 14% of the US Population.[34] Even within households, emotional labor and care remain "women's" work (Erickson 2005; Atanasoski and Vora 2019).

The stay-at-home-mom is the paradigm of care labor that a capitalist system expects sans monetary recompense, without which much of our economy would not function (See: Cowan 1983; Antonopoulos 2008). Care work jobs are paid less than many other, perhaps less-necessary, professions, and are often near or at minimum wage (Atanasoski and Vora 2019; Antonopoulos 2008). This produces a set of workers who are stressed and precarious, lowering the level of care they are able to produce. Then, those dependent upon them, those who are most in need of care, often get poor quality care, or care that is fractured because of either care workers leaving the field because it does not support them, or because they must shop around for new care workers if theirs are insufficiently trained or capable. This tension can create mechanisms for abuse and violence in both directions. Either from care worker to client (the annual disability day of mourning for disabled people murdered by their family and caretakers is an especially tragic example), or from the client to a care worker who may not have the recourse to refuse a client or to change jobs. This can be due to a limited possible clientele, the immigration status of the care worker, a fear of losing income and thus becoming housing or food insecure, or any other number of possible issues.

This devalued set of labors is central to how the communities I described in this chapter live and thrive. I argue that care must be the central motivating social force of a morphological freedom that actively values morphological difference. The kinds of care networks that biohackers, body modders, Trans and queer people, and Disabled people create and maintain already hold up their alternative morphologies as valuable and desirable. In order for morphological freedom to become the human right that transhumanists desire, care *must* be the mechanism by which it is established.

9.7 Conclusion

I mentioned in the introduction of this chapter that there are three steps that I want to focus on to produce a better morphological freedom. Eliminating (or at least reducing the impact of) capitalism, valuing care and care work, and dedicating ourselves to the value of intra-dependence. These three steps are all part of producing a world which can actually value morphological difference on a large scale. They are not the only things we would need to do, and it's not guaranteed that if we succeed in these three steps that valuing morphological difference would necessarily result. However, I doubt any world in which these three steps aren't taken could adequately value morphological difference in the way that we ought to want.

---

[34] Numbers compiled from US Census and demographic data found here: https://www.cen sus.gov/newsroom/releases/archives/2010_census/cb11-cn125.html; https://www.bls.gov/cps/cps aat18.htm.

Current events as I write this, namely the SARS-CoV2 (COVID-19) outbreak, have also brought into stark relief the way society could be... if only we chose to make it so. Universal Basic Income is being rolled out in several countries. Food delivery options which disabled people often rely upon are showing their potential value to us all. The moniker of "essential worker" is similarly being shown to be mostly those laborers we (economically and socially) value least: fast food, grocery workers, warehouse workers, shipping and transportation, health care, waste disposal, cleaning and housekeeping, etc. Many of those essential workers are beginning to strike for benefits and protections and pay equal to their value. Poor and middle-class people are coming together in mutual aid networks (see: Spade 2020) to support each other, while the richest people in the country fight against giving their workers bathroom breaks or union protection. The exploitation of capitalism is beginning to be visible to more than just the Marxist activists in internet chatrooms and academic ivory towers.

In order to get out of the neoliberal mindset, and into a mindset that values care, intra-dependence, mutual aid, and multiple morphologies, we must take this opportunity to move the political needle. Expand access to health care, pay at-home care workers what they are worth, and value those folks who break out of our normative idea of what a bodymind should be. Disabled people and communities, trans people and communities, and body modders all have a lot of experience giving and receiving care (biohackers perhaps less so), and can teach us a lot about what is necessary. All of these communities have been excluded from society at large. All have found themselves unable to fully participate in the workforce and thus are refused the category of the neoliberal subject. All of them understand what kind of caring and intra-dependent[35] society there could be. Many are adept at producing and maintaining care networks, and mutual aid systems with friends, family, neighbors, and other marginalized groups (Piepzna-Samarasinha, 2018; Spade 2020). Another thing that the recent pandemic has brought to light is how necessary interaction, and even simple touch, can be. These practices of being together, touching, conversing, sharing moments, are all a part of care, are all a part of intra-dependence, and are all anathematic to neoliberal individualism.

In this chapter I have described how the experience of those with non-normative bodyminds troubles the neat assumptions of transhumanists about morphological freedom. I also argue that they can teach us how better to establish a morphological freedom that actually does what transhumanists claim to want.[36] The twofold

---

[35] "Intra-dependent" pulls from Barad's (2007) "intra-action," which argues that actions between two agents cannot be considered simply transactional, but rather that they change each actor, and also move beyond individual actors to alter the networks that hold those actors up (cf Latour). She calls this intra-action (as opposed to interaction). Thus, I claim that dependence—via care as the intra-actional mechanism—works the same; it is not a transactional relationship between two individuals, but rather a widespread community endeavor that both requires large networks, but also changes them through their actions. Further explication of this can be found in the conclusory chapter of my dissertation (Earle 2021b).

[36] I argue elsewhere that the claim they make about what morphological freedom is and will do is not made in good faith, and is instead a cover for other goals (Earle 2021a, b).

approach I argue we need to take, includes the work of embracing care that I have already referenced throughout this chapter, but also an idea that I have been developing, expanding on work by Barad (2007), Haraway (1991, 1992, 2016), and Crenshaw (1989, 1991) and informed by the work on care already referenced. I call this idea radical intra-dependence. Intra-dependence is the idea that we are not only dependent upon the people and things which surround us and allow us to live day to day, but dependent *through* them into the networks which hold those things and people up as well. It is not just the house or apartment that we depend on to protect us from the elements, but the lumber, construction, and maintenance companies necessary to build those houses. We are also dependent upon more people than we might immediately think about. We may depend on our care worker, but we depend through them on their trainer, their family and support system. And in turn they depend on us and through us to our own support systems. The house cannot get built without someone willing to buy it who has a support system that helps them do so. Intra-dependence is not a transactional relationship, but an ontological one. We cannot exist without our relations, without our dependencies. In short: we are all dependent on and through others[37] to such a degree that notion of the individual itself is a fiction. A useful fiction, perhaps, particularly when describing phenomenological experience, oppression, and positionality against power structures, but a fiction nonetheless.

The key takeaway from this notion of radical intra-dependence and the absence of the individual is that multiplicity becomes a foundational generative force: the main mechanism by which new possibilities/ways of being are produced. Through the intra-action of difference, new possibilities emerge. On one hand, the claim that multiplicity is generative may seem trivial. Obviously, static homogenous organizations will produce static, homogenous things more often than not. Longino (1990) has made this argument for diversifying scientific labs. Through introducing multiple perspectives, biases held by majority populations (read: white men) are more likely to be noticed and removed from the science. This is, in effect, protecting against the weakness that comes from breeding a monoculture. Haraway (1991)—in similar fashion to Longino—describes how partial perspectives lead to a better objectivity, one that is not deceived into thinking it can see from everywhere and nowhere simultaneously (the "god trick").

To this end, care is the mechanism by which those generative relationships are nurtured. Care has been a relatively recent concept about which to theorize in STS, though it is, of course, foundational to practices which go back millennia (nursing, midwifery, etc.). A foundational work on care within STS was 2008s *Logics of Care* by Annemarie Mol. Her prior book, *The Body Multiple* (2002), with its expansive take on medical practice (including, tangentially, some care practices), begins to break down the notion that ontology is a singular thing. Instead, as a thing (in the case of her book, atherosclerosis) moves from one context to another, ontologies multiply. The phenomenological ontology of pain while walking is not the same

---

[37] There is a good deal of Indigenous scholarship on kinmaking and dependency that I know exists, but have not yet been able to engage. That work is important here, but I cannot do it enough justice to include at this time. I look forward to integrating it into my own work soon.

as the ontology of scleral thickening under a microscope, or the dropping of blood pressure in the clinic. And, in fact, each of these ontologies requires multiple *actant*[38] in order to be produced. The atherosclerosis of a patient requires a doctor, their attendant hospital and medical system, labs, chemicals, microscopes, etc. to exist at all, and exists differently depending on which perspective you take and which arrangement of *actant* are currently producing the atherosclerosis. The notion that such a thing as atherosclerosis cannot exist outside of that intra-action, but a person can, becomes absurd on its face. We cannot exist outside of our intra-actions, outside of our intra-dependencies. We cannot be considered, ethically, epistemologically, or ontologically, individuals. If you strip everyone else away, we cease to be.

Because we are intra-dependent, dependency itself must be destigmatized. We must see it for the foundational piece of the human experience that it is, and that care is its mechanism. By focusing on the contested and stigmatized notion of dependence, viewing it through a Baradian diffractive lens to see how we depend not *on* each other, but *through* each other, I lean into the conflicting and incommensurable ways in which care both resists and reinforces problematic views on the bodymind. And in keeping with Eli Clare and Ewa Ziarek, I refuse the notion that a single ethic can capture and correct the manifold violences that occur, but rather sit within what Ziarek calls an "ethics of dissensus," where the incommensurability itself becomes generative of new mechanisms of resistance and agency (Ziarek 2001).

As Maria Puig de la Bellasca writes: "To care can feel good; it can also feel awful. It can do good; it can oppress…Care is not about fusion; it can be about the right distance… It also doesn't mean that *to care* should be a moral obligation in all situations, practices, or decisions" (2017, pp. 1 and 5, emphasis in original). In order to get to an ethical, possible, and thriving state of morphological freedom, the transhumanist community ought to accept our intra-connectedness, our intra-dependence, and work on fostering a mechanism of social care, which can help morphologically diverse communities thrive. Of course, there are barriers, not the least of which is a focus within transhumanism, but also within everyday society, that an individualist bootstrappy way of life is the natural way for us all to be. Thus, it becomes incumbent upon the rest of us, who may not be of the same mind as transhumanists, who may not have ever considered how we might alter our bodyminds if given the opportunity, to push back against a neoliberal production model of individual value, and toward a more communitarian system. Once we all (or at least a good plurality) can value difference, can see the generative power of multiplicity, both in bodyminds, but also in communities, we might be able to have a morphological freedom that enhances our connectivity rather than our separateness. In turn, a lot of communities will be able to come in from the margins of society, and teach us about the care, intra-dependence, and meaning-making that they use to produce thriving communities toward which we all might aspire.

---

[38] "*Actant*" are nodes in an actor-network and can be human or non-human (see: Latour 1987). The use of *actant*, here, instead of "actor" foregrounds the presence and importance of the nonhuman parts of the actor-network.

References

Adams, R., B. Reiss, and D. Serlin. 2015. *Keywords for disability studies*. EBL-Schweitzer. NYU Press.

Antonopoulos, Rania. 2008. The unpaid care work-paid work connection. *Levy Economics Institute, Working Papers Series*.

Atanasoski, N., and K. Vora. 2019. *Surrogate humanity: Race, robots, and the politics of technological futures*. Perverse modernities: A Series Edited by Jack Halberstam and Lisa Lowe. Durham: Duke University Press.

Bacon, Francis. 1597. *Sacred meditations*.

Bangh, Aimee. 2018. *Migrant futures: Decolonizing speculation in financial times*. Durham: Duke University Press.

Barad, Karen. 2007. *Meeting the universe halfway: Quantum physics and the entanglement of matter and meaning*. Durham: Duke University Press.

Bercovici, Jeff. 2016. Peter Thiel is very, very interested in young people's blood. *Inc.*, August 2016. https://www.inc.com/jeff-bercovici/peter-thiel-young-blood.html.

Brashear, Regan. 2014. *Fixed: the science/fiction of human enhancement*.

Brown, Kristen. 2017. The FDA is not cool with selling DIY gene therapies. *Gizmodo*, November 2017. https://gizmodo.com/the-feds-are-officially-cracking-down-on-basement-bioha-182068 2025.

Brown, Kristen. 2018. What does an infamous biohacker's death mean for the future of DIY science? *The Atlantic*, May 2018. https://www.theatlantic.com/science/archive/2018/05/aaron-traywick-death-ascendance-biomedical/559745/.

Clare, E. 2015. *Exile and pride: Disability, queerness, and liberation*. Durham: Duke University Press.

Clare, E. 2017. *Brilliant imperfection: Grappling with cure*. Duke University Press.

Clark, Terryann C., Lucassen, M.F.G, Bullen, P., Denny, S.J., Fleming, T.M., Robinson, E.M., and Fiona V. Rossen. 2014. The health and well-being of transgender high school students: Results from the New Zealand Adolescent Health Survey (Youth'12). *Journal of Adolescent Health* 55 (1): 93–99. https://doi.org/10.1016/j.jadohealth.2013.11.008.

Collins, P. H. 2000. *Black feminist thought: Knowledge, consciousness, and the politics of empowerment*. Perspectives on gender. New York & London: Routledge.

Cowan, Ruth Schwartz. 1983. *More work for mother: The ironies of household technology from the open hearth to the microwave*, vol. 5131. New York: Basic Books.

Crenshaw, Kimberle. 1989. Demarginalizing the intersection of race and sex: A black feminist critique of antidiscrimination doctrine, feminist theory and antiracist politics. *University of Chicago Legal Forum* 139.

Crenshaw, Kimberle. 1991. Mapping the margins: Intersectionality, identity politics, and violence against women of color. *Stanford Law Review* 1241–99.

de la Bellacasa, Maria Puig. 2017. *Matters of care: Speculative ethics in more than human worlds*, vol. 41. Minneapolis: U of Minnesota Press.

Delfanti, A. 2013. *Biohackers: The politics of open science*. London: Pluto Press.

DeMello, M., and G.S. Rubin. 2000. *Bodies of inscription: A cultural history of the modern tattoo community*. Durham: Duke University Press.

Earle, Joshua. 2021a. Engineering Our Selves: Morphological freedom and the myth of multiplicity. In *Philosophy and engineering: Reimagining technology and social progress*, ed. Zachary Pirtle, Guru Madhavan, and David Tomblin, 249–267. New York: Springer.

Earle, Joshua. 2021b. *Morphological freedom and the construction of bodymind malleability from eugenics to transhumanism*. https://vtechworks.lib.vt.edu/handle/10919/107009.

Elliott, Carl. 2003. *Better than well: American medicine meets the American dream*. New York: WW Norton & Company.

Elliott, Carl. 2004. *Better than well: American medicine meets the American dream*. WW Norton & Company.

Epstein, Steven. 1995. The construction of lay expertise: AIDS activism and the forging of credibility in the reform of clinical trials. *Science, Technology, & Human Values* 20 (4): 408–437.

Erickson, Rebecca J. 2005. Why emotion work matters: Sex, gender, and the division of household labor. *Journal of Marriage and Family* 67 (2): 337–351.

Forestiere, Annamarie. 2020. America's war on black trans women. *Harvard Civil Rights—Civil Liberties Law Review*, September 2020. https://harvardcrcl.org/americas-war-on-black-trans-women/.

Fuller, Steve. 2016. Morphological freedom and the question of responsibility and representation in transhumanism. *Social Epistemology*, 33–45.

Garland-Thomson, Rosemarie. 2002. Integrating disability, transforming feminist theory. *NWSA Journal* 1–32.

Gilbert, Melody. 2003. *Whole*.

Haraway, D.J. 1991. *Simians, cyborgs, and women: The reinvention of nature. Simians, cyborgs, and women*.

Haraway, Donna. 1992. The promises of monsters: A regenerative politics for inappropriate/d others. *Cultural Studies* 295–337.

Haraway, D.J. 2016. *Staying with the trouble: Making kin in the Chthulucene. Experimental futures*. Durham: Duke University Press.

Hellstrom, Per M. 2006. This year's Nobel Prize to gastroenterology: Robin Warren and Barry Marshall awarded for their discovery of Helicobacter Pylori as pathogen in the gastrointestinal tract. *World Journal of Gastroenterology* 12 (19): 3126–3127. https://doi.org/10.3748/wjg.v12.i19.3126.

Herdegen, Hanna. 2019a. Maintaining disabled bodies and identities: disability as dirty work. *The Maintainers*. 2019. http://themaintainers.org/blog/2019a/6/17/maintaining-disabled-bodies-and-identities-disability-as-dirty-work.

Herdegen, Hanna. 2019b. Maintaining disabled bodies and identities: the body as evidence. *The Maintainers*. 2019. http://themaintainers.org/blog/2019b/6/21/maintaining-disabled-bodies-and-identities-the-body-as-evidence.

Isaacson, Walter. 2021. *The code breaker: Jennifer Doudna, gene editing, and the future of the human race*. New York: Simon & Schuster.

Istvan, Zoltan. 2015. In the transhumanist age, we should be repairing disabilities, not sidewalks. *Motherboard*, April 2015. https://motherboard.vice.com/en_us/article/4x3pdm/in-the-transhumanist-age-we-should-be-repairing-disabilities-not-sidewalks.

Jackson, Liz. 2019. A community response to a #DisabilityDongle. *The inclusive Liz Jackson*. 2019. https://medium.com/@eejackson/a-community-response-to-a-disabilitydongle-d0a37703d7c2.

Juno, A., and V. Vale. 1989. *Modern primitives: An investigation of contemporary adornment & ritual*. California: Re/Search Publications.

Kant, Immanuel. 1781. *Critique of pure reason*.

Kasnitz, Devva. 2020. The politics of disability performativity: An autoethnography. *Current Anthropology* 61 (S21): S16-25.

Latour, Bruno. 1987. *Science in action: How to follow scientists and engineers through society*. Harvard University Press.

Lawrence, Anne A. 2004. Autogynephilia: A paraphilic model of gender identity disorder. *Journal of Gay & Lesbian Psychotherapy* 8 (1–2): 69–87.

Longino, Helen E. 1990. Values and objectivity. In *Philosophy of science: The central issues*, ed. Martin Curd and J.A. Cover, 62–82. Princeton: Princeton University Press.

Metz, Rachel. 2019. Elon Musk hopes to put a computer chip in your brain. Who wants one? *CNN Business*, 21 July 2019. https://www.cnn.com/2019/07/20/tech/elon-musk-neuralink-brain-chip-experts/index.html.

Mol, Annemarie. 2002. *The body multiple: Ontology in medical practice*. Durham: Duke University Press.

Mol, Annemarie. 2008. *The logic of care: Health and the problem of patient choice*. New York: Taylor & Francis.

More, Max, and Natasha Vita-More. 2013. *The transhumanist reader: Classical and contemporary essays on the science, technology, and philosophy of the human future.* New York: Wiley.

O'Toole, C.J., K. Nakamura, and E. Grace. 2015. *Fading scars: My queer disability history.* Texas: Autonomous Press.

Paine, Thomas. 1794. *The age of reason; being an investigation of true and fabulous theology.*

Paine, Thomas. 1796. *The age of reason. Part the Second ...* 2nd ed. D. I. Eaton.

Paine, Thomas. 1811. *The age of reason: Part the Third : Being an examination of the passages in the new testament ...* to Which Is Prefixed an Essay on Dream ... with an Appendix Containing My Private Thoughts of a Future State. Printed, published and sold by Daniel Isaac Eaton.

Party, Transhumanist. 2018. *Transhumanist Bill of Rights—Version 3.0.* 2018. https://transhuma nist-party.org/tbr-3/.

Piepzna-Samarasinha, L.L. 2018. *Care work: Dreaming disability justice.* Vancouver: Arsenal Pulp Press.

Pitts, V. 2003. *In the flesh: The cultural politics of body modification.* New York: Palgrave Macmillan US.

Preciado, P.B., and B. Benderson. 2013. *Testo junkie: Sex, drugs, and biopolitics in the pharmacopornographic era.* New York: Feminist Press at CUNY.

Price, Margaret. 2015. The bodymind problem and the possibilities of pain. *Hypatia* 30 (1): 268–284.

Puar, J.K. 2017. *The right to maim: Debility, capacity, disability.* ANIMA. Durham: Duke University Press.

Reardon, Sara. 2015. Epigenetic 'tags' linked to homosexuality in men. *Nature News*, October 2015.

Rice, George, Carol Anderson, Neil Risch, and George Ebers. 1999. Male homosexuality: Absence of linkage to microsatellite markers at Xq28. *Science* 284 (5414): 665–667.

Rothblatt, Martine. 2011. *From transgender to transhuman: A manifesto on the freedom of form.* New York: Princeton Architectural Press.

Samuels, Ellen. 2017. Six ways of looking at crip time. *Disability Studies Quarterly* 37 (3). https://dsq-sds.org/article/view/5824/4684.

Savulescu, Julian. 2003. Human-animal transgenesis and chimeras might be an expression of our humanity. *The American Journal of Bioethics* 3 (3): 22–25.

Savulescu, Julian, and Guy Kahane. 2011. Disability: A welfarist approach. *Clinical Ethics* 6 (1): 45–51. https://doi.org/10.1258/ce.2011.011010.

Schalk, Sami. 2018. *Bodyminds reimagined: (Dis)ability, race, and gender in black women's speculative fiction.* Duke University Press.

Selyukh, Alina. 2020. Workers with disabilities can earn just $3.34 an hour, agency says law needs change. *NPR Radio.* 2020. https://www.npr.org/2020/09/17/912840482/u-s-agency-urges-end-to-below-minimum-wage-for-workers-with-disabilities.

Spade, Dean. 2020. *Mutual aid: Building solidarity during this crisis (and the Next).* Verso Books.

Thorsen, Dag Einar. 2010. The neoliberal challenge—What is neoliberalism. *Contemporary Readings in Law and Social Justice* 2 (2): 188–214.

Turner, William J. 1995. Homosexuality, Type 1: An Xq28 phenomenon. *Archives of Sexual Behavior* 24 (2): 109–134.

Weise, Jillian. 2016. The dawn of the Tryborg. *The New York Times*, 30 November 2016. https://www.nytimes.com/2016/11/30/opinion/the-dawn-of-the-tryborg.html.

Wittkower, Dylan Eric. 2016. Lurkers, creepers, and virtuous interactivity: From property rights to consent to care as a conceptual basis for privacy concerns and information ethics. *First Monday* 21 (10).

Wittkower, Dylan Eric. 2020. Privacy as care: An interpersonal model of privacy exemplified by five cases in the internet of things. In *Relating to things: Design, technology and the artificial*, ed. Heather Wiltse. London: Bloomsbury Publishing.

Wong, Alice. 2020. Disabled oracles and the coronoavirus. *Disability Visibility Project.* 2020. https://disabilityvisibilityproject.com/2020/03/18/coronavirus/.

Worthen, M.G.F. 2016. *Sexual deviance and society: A sociological examination.* New York: Taylor & Francis.

Young, Stella. 2014. I'm not your inspiration, thank you very much. *TED*. 2014. https://www.ted.com/talks/stella_young_i_m_not_your_inspiration_thank_you_very_much.

Ziarek, Ewa P. 2001. *An ethics of dissensus: Postmodernity, feminism, and the politics of radical democracy*. Stanford University Press.

Dr. Joshua Earle is a scholar and researcher who focuses on the political and ethical ramifications of technology and medicine, particularly that which is supposed to "enhance" human body-minds. He received his Ph.D. in Science and Technology Studies as well as a Graduate Certificate in Women and Gender Studies from Virginia Tech. His work explicitly engages Critical Race Studies, Disability Studies, Queer Theory, and (Black) Feminist Theory. Send any inquiries to jearle@vt.edu.

Chapter 10
Posthuman Ethics: The Priority of Ethical Over Ontological Status

Sanja Ivic

Abstract The moral status of new beings has often been linked to the question of the ontological status of new beings. Some authors believe that the discussion of the ontological status of new beings should be abandoned, in order to give priority to ethics in light of Lévinas's idea that ethics should take precedence over ontology. This chapter shows that the question of the subjectivity of new beings cannot be avoided within legal and ethical considerations. The postmodern theory offers an adequate theoretical framework for the development of transhuman and posthuman ethics. Transhuman agency requires postmodernist understanding of subjectivity, which leaves room for hybrid identities. On the other hand, the posthuman ethics requires the postmodern understanding of morality which accepts otherness and overcomes binary hierarchies based on power relations.

10.1 Introduction

The denialism of non-human subjectivity and agency stems from the early modern notion of the self. The philosophers of the Enlightenment universalized human beings by excluding and marginalizing all other beings. This idea of humanity stems from the idea of subjectivity mostly defined by philosophers René Descartes (1993) and John Locke (1836). The Cartesian notion of the self is described as pure, unified, conscious and rational. Descartes defined the subject as a "thinking thing": "But what then am I? A thing that thinks. What is that? A thing that doubts, understands, affirms, denies, is willing, is unwilling, and also imagines and has sensory perceptions" (Descartes 1993, 19). According to John Locke, subjectivity may be ascribed to a "thinking intelligent being" endowed with reason (Locke 1836). In this way a sharp distinction is based between the subject and object, human and non-human, organic and mechanic and so forth. Both Descartes's and Locke's philosophy of subjectivity reflect denial of non-human subjectivity, which is perceived as the Other. According

S. Ivic (✉)
Institute for European Studies, Belgrade, Serbia
e-mail: sanja_ivic1@yahoo.com

to Gunkel, the concept of the human has always been exclusionary. "It has always been a way of excluding others, whether they were African peoples, whether they were Aboriginal peoples, whether they're women, whether they are animals, there is a way in which the concept of the human has been a way for one group in power to disempower others" (Gunkel 2014).

Postmodern and poststructuralist theorists[1] argue that the idea of identity developed by the Enlightenment thinkers is flawed (St. Pierre 2000). The Enlightenment thinkers perceive the concept of the self as natural and fixed and derive truth and knowledge from this notion of the self. Representatives of poststructural[2] and postmodern theory perceive subjectivity as performative, not constative (Derrida 1986). Consequently, subjectivity does not pre-exist the discursive field. This means that "we come to understand who we are through the re-iteration or performance of identity. In this sense, identity is not about fixed attributes possessed by individuals, but is instead constructed in a variety of ways at a variety of levels" (Morgan 2000, 217). The postmodern idea of subjectivity is polyphonic and fluid and leaves room for the Other. The modern conception of the self is based on the dualism between humanity and nature—between subject and object. Postmodern philosophy aims at reconstructing and rethinking these dualisms.

This chapter will argue that postmodern ethics is necessary for the development of posthuman ethics (Pruchnic 2013). Philosophical theories labeled as postmodernism, poststructuralism and deconstruction have a significant role in developing posthuman ethics.Postmodern ethics is based on the idea of "responsibility to otherness" (Gibbons 1991, 96).

The postmodern idea of the self unites human and non-human (mechanical, animal, environmental, and so forth). Transhumanism developed during the postmodern era and shares some postmodernist perspectives and values, such as the idea of a polyphonic, multilayered notion of identity, reevaluation of knowledge, and rejecting sharp binary opposition between the human and non-human (More and Vita-More 2013, 1).

---

[1] According to Butler, some characterizations "are variously imputed to postmodernism and post-structuralism, which are conflated with each other and sometimes conflated with deconstruction, and sometimes understood as an indiscriminate assemblage of French feminism, deconstruction, Lacanian psychoanalysis, Foucaultian analysis, Rorty's conversationalism and cultural studies. On this side of the Atlantic and in recent discourse, the terms 'postmodernism' or 'poststructuralism' settle the differences among those positions in a single stroke, providing a substantive, a noun, that includes those positions as so many of its modalities or permutations. It may come as a surprise to some purveyors of the Continental scene to learn that Lacanian psychoanalysis in France positions itself officially against post-structuralism, that Kristeva denounces postmodernism, that Foucaultians rarely relate to Derrideans, that Cixous and Irigaray are fundamentally opposed, and that the only tenuous connection between French feminism and deconstruction exists between Cixous and Derrida, although a certain affinity in textual practices is to be found between Derrida and Irigaray"(Butler 2001, 630).

[2] "Rather than viewing self as an objectifiable, cognitive essence, poststructuralists argue that identity processes are fundamentally ambiguous and always in a state of flux and reconstruction" (Collinson 2006, 182).

The notions of transhuman and posthuman are often used interchangeably, despite representing two different concepts. Although transhumanism has developed in the postmodern era, it has its origins in the Enlightenment, which sought to master nature and technologies. Transhumanism strives for the postbiological evolution of the human body and mind with the ultimate goal of overcoming the limited human condition (Koljenik 2014). One of the most important theorists of transhumanism, Nick Bostrom, defines a posthuman being as follows: "Transhumanists hope that by responsible use of science, technology, and other rational means we shall eventually manage to become posthuman, beings with vastly greater capacities than present human beings have" (Bostrom 2003).

Transhumanism represents a project of continuous modification of human beings (Manzocco 2019). However, this project remains anthropocentric, although it rethinks the notion of anthropocentrism and leads toward "techno-anthropocentrism" (Thomas 2017). On the other hand, the term posthuman refers to transformation of ideals of humanity. In this way, it overcomes anthropocentrism as it understands the concept of human as a socially, culturally, and historically constructed term. According to Hayles, "the posthuman subject is an amalgam, a collection of hetero-geneous components, a material-informational entity whose boundaries undergo continuous construction and reconstruction" (Hayles 1999, 3). While transhumanism focuses on creating the posthuman condition by perfecting and upgrading the body (which is related to the concept of the cyborg), posthumanism is radically moving away from the body in order to create a posthuman state of being embodied in an immortal virtual identity, artificial intelligence or robot (Koljenik 2014). However, they overlap in many ways, and sometimes it is not easy to strictly delineate them (Pilsch 2017). The reason for this is that transhumanist philosophy also contains a final vision of abandoning the limitations of the biological body and is interested in similar topics as techno-scientific posthumanism (immortality, artificial intelligence, and so forth) (Koljenik 2014).

Posthumanism includes different approaches. Critical posthumanism and specu-lative posthumanism represent the most common approaches (Roden 2014). Both approaches reject anthropocentric perspectives. According to Roden, "whereas crit-ical posthumanists are interested in the posthuman as a cultural and political condi-tion, speculative posthumanists are interested in a possibility of certain technologi-cally created things" (Roden 2014). On the other hand, transhumanism shares many aspects of humanism, such as a commitment to progress and respect for science and reason (Philosophy Question 2019). "Posthuman" describes the state in which humans have ceased being human, whereas "transhuman" is the transitionary state between human and posthuman, in which evolution is achieved via human enhance-ment (i.e. nexus of technoscience, body augmentation, and spiritual enlightenment)" (Philosophy Question 2019).

Koljenik (2014) distinguishes between two forms of posthumanism: 1. techno-scientific posthumanism, which sees overcoming the limited human condition in radical abandonment of the body and devising a new virtual identity, new forms of artificial intelligence that will overcome the limitations of the biological body through virtual immortality, and 2. philosophical posthumanism; which arose within postmodernist philosophy as an attempt to break the link with humanism and anthropocentrism, as well as all the dualisms and determinisms they imply.

Both proponents of transhumanism, on the one hand, and posthumanism, on the other, aim at rethinking the rationalist tradition of Enlightenment humanism (Hauskeller 2016). However, transhumanism seeks to expand the aspirations of the Enlightenment (Levin 2021) and humanism, while posthumanism transcends them.

10.2   The Question of Subjectivity and Moral Status of Novel Beings

The contemporary era is based on the following dualisms: human/non-human, labor/robotics, mind/artificial intelligence, reproduction/replication, representation/simulation, and so forth (Haraway 2016, 28–29). "Certain dualisms have been persistent in Western traditions (…) self/other, mind/body, culture/nature, male/female, civilized/primitive, reality/appearance, whole/part, agent/resource, maker/made, active/passive, right/wrong, truth/illusion, total/partial, God/man" (Haraway 2016, 59–60). The theorists of posthumanism, often use the concept of posthuman as a metaphor which aims at deconstructing the basic dualisms of the Western rationalist tradition, and especially that between subject and object (Herbrechter 2021). They aim at reconceptualizing and transforming the modernist Enlightenment notion of the subject as unified, coherent, autonomous, and rational. On the other hand, "in deconstruction and poststructuralism, humanist narratives of progressive self-understanding and mastery are challenged by ontologies which, like cyborg theory, resist any description of the human subject as a self-present source of meaning or self-authenticating source of value" (Roden 2010, 31).

Donna Haraway (1987) employs the figure of cyborg in order to criticize essentialist notions of identities that are considered fixed and natural. According to Haraway, the figure of cyborg, based on hybrid identity, has huge deconstructive potential (Haraway 1987). A cyborg represents fluid and heterogeneous identities, transcending fixed boundaries—those separating human from non-human (Haraway 1987). Harraway aims at deconstructing the modernist notion of fixed, homogenous identity and the power relations on which it is based so that this newly established cyborg-ontology would give birth to a new policy. Donna Haraway uses the concept of cyborg as one of the forms of postmodern rethinking of the Enlightenment narratives of subjectivity and patriarchal values. On the other hand, she uses this concept as an expression of hope into the possibility of some different epistemology and ontology in which identity would be hybrid, never absolutely present, where biological or cultural identity would not limit opportunities and existences, where the body would freely merge with other prosthetic means and other bodies, outside binary oppositions, where power would not be equated with hierarchical relations, and a different epistemology would create a different policy, which would not aim to fight for borders (Zivkovic 2012).

The posthuman era opens the space for the development of various hybrid entities and identities that transcend binary oppositions: human/machine, human/animal, labor/robotics, and so on (Haraway 2016, 28–29). The posthumanist notion of hybrid and polyphonic identities enables pluralism of votes and rights, including animals, plants, and the environment who may be treated as subjects of law.

In light of the cyborg, the concept of the "human" needs rethinking (Hassan 1977). The development of technology has opened up a space for reconceptualizing the concepts of the organism, on the one hand, and the machine, on the other. "The ontology grounding 'Western epistemology'" (Haraway 2016, 12) needs to be transformed as well. The question of subjectivity is significant in discussions regarding the moral status of novel beings. However, it seems that insights (regarding the contemporary notions of identity) from critical posthumanism, poststructuralism, and postmodernism are not sufficiently taken into account.

One group of authors who discuss the moral status of novel beings rely on the argument called "Chinese room", presented by the philosopher John R. Searle (1980). Searle presented the Chinese room argument as a thought experiment to demonstrate the impossibility of explaining the structure and function of human cognition by identifying it with the structure and function of computers. According to Searle, the essential difference between man and computer is that man, unlike computers, understands the meaning of words, and therefore any hope of constructing strong artificial intelligence is futile. Searle argues:

> … The point of the story is simply this: by virtue of implementing a formal computer program from the point of view of an outside observer, you behave exactly as if you understood Chinese, but all the same you don't understand a word of Chinese (…) To repeat, a computer has a syntax, but no semantics. The whole point of the parable of the Chinese room is to remind us of a fact that we knew all along. Understanding a language, or indeed, having mental states at all, involves more than just having a bunch of formal symbols. It involves having an interpretation, or a meaning attached to those symbols. And a digital computer, as defined, cannot have more than just formal symbols because the operation of the computer, as I said earlier, is defined in terms of its ability to implement programs. And these programs are purely formally specifiable – that is, they have no semantic content (Searle 1984, 32–33).

Searle (1980) emphasizes that a computer program doesn't understand the symbols it manipulates. Thus, "syntax alone is not sufficient for semantics." (Searle 2002, 672). The proponents of this point of view, extend Searle's argument to all kinds of novel beings—cyborgs, robots, and so on. According to this point of view, these novel beings cannot be considered moral agents.

Critics of Searle's arguments point to the "other minds problem", and emphasize that consciousness is a private experience which cannot be measured by objective criteria (Gunkel 2018). According to Gunkel, "not only are these efforts unable to demonstrate with any certitude whether animals, machines, or other entities are in fact conscious (…) and therefore legitimate moral persons (or not), we are left doubting whether we can say the same for other human beings" (Gunkel 2018, 93).

Some authors argue that "an isomorphic processing system made entirely of non-organic materials that runs similar [brain] software would be conscious—regardless of whether it is fairly characterized as 'biologically alive'" (Himma 2009, 26).

Terminological complications represent another problem, as a unified definition of consciousness does not exist, and this term is differently interpreted by philosophers, cognitive scientists, anthropologists, neurobiologists, and so forth (Gunkel 2018, 92). David Gunkel argues that an agent-oriented problematic regarding the ethics of robots and artificial intelligence needs to "shift the focus and consider things from the other side—the side of machine moral patiency" (Gunkel 2018, 87). The machine question poses a profound challenge to moral reasoning, calling into question the traditional philosophical understanding of technology as a tool or instrument to be used by human agents (Gunkel 2012). Gunkel (2012) discusses the issue of machine moral agency, or whether a machine can be considered a valid moral agent capable of being held accountable for its decisions and acts. He also investigates whether a machine may be a moral patient due valid moral consideration. Moral patients are those who are to be the subject of moral consideration and deserve it, although they lack agency, such as animals, allowing moral consideration to be wider and more inclusive (Denton 2014). Gunkel (2012) examines some recent developments in moral philosophy and critical theory that complicate the machine debate, deconstructing the binary opposition agent/patient.

This idea of moral patiency is also emphasized by cyborg activists. Cyborg-rights activists propose cyborg rights. Activist Aral Balkan drafted the Universal Declaration of Cyborg Rights (2017). The purpose of this declaration is to extend the notion of the self in light of developing technologies. T This declaration aims at extending the scope of the Universal Declaration of Human Rights (1948). However, the draft of this declaration is not finished yet. The drafters of the Universal Declaration of Cyborg Rights argue:

> No nation has a constitutional decree that defines the technologies that we use to extend our minds (and thus selves) with as existing within the boundaries of the self. In short, our constitutional definition of personhood today does not include all aspects that make up the person, and thereby the laws we have to protect the rights of people fall short in protecting them fully. The sole purpose of The Universal Declaration of Cyborg Rights is to hack/patch The Universal Declaration of Human Rights to extend the scope of what constitutes a person in the era of networked digital technology so that we can ensure that the existing rights we already have cover the entirely of the self of what constitutes a person in our age (Aral Balkan 2017).

Thus, the question of moral rights and duties is tied to the question of subjectivity. The goal of this declaration is not to propose a new set of rights, but to extend the notion of the self on which rights are based, in order to include technologies that extend the concept of the self (Aral Balkan 2017).

Article 1 of the Universal Declaration of Cyborg Rights states that "human beings in the digital age are cyborgs; sharded beings" (Aral Balkan 2017). As it is already mentioned, the main goal of this declaration is to extend the scope of the Universal Declaration of Human Rights (UDHR). However, the UDHR includes a number of vague terms, which point to various interpretations.

> But philosophers disagree on how to define human dignity and, as with human rights, the concept is often regarded as a Western one not applicable to other cultures. On the other hand, with the recognition of poverty and climate change as major violations of human rights

and faced with certain challenges to the uniqueness of humanity caused by modern science and technology, notably biomedicine and genetic engineering, the concept of human dignity features again more prominently in the contemporary human rights discourse (Nowak 2001, p. v).

Article 3 of the Universal Declaration of Cyborg Rights aims at protecting "integrity and dignity of the cyborg self" (Aral Balkan 2017). But even the concept of dignity of human beings is still not well defined (Howard and Donelly 1986; Macklin 2003; Schroeder 2012), and if cyborgs represent the extended self, the same definition of dignity applies to cyborgs.

The Western conception of legal subjects and rights is derived from the philosophy of the Enlightenment. This philosophy represents a metaphysical and ontological foundation of the Universal Declaration of Human Rights (1948). Thus, the notion of rights is tied to traditional moral philosophy which assigns rights to human beings perceived as rational and conscious beings. Recently, this point of view began to change with the development of animal and environmental rights. However, these new subjects of rights are still perceived as Other.

The development of the rights of cyborgs and other novel beings depends on the development of the main concepts on which human rights are based (such as personhood, dignity, integrity, and soforth). "Cyborg citizenship is a conception of rights based on personhood rather than on 'humanness'. Not all persons are humans, and not all humans are persons" (Cyborg Citizenship n.d.). However, the question then becomes what does personhood mean. This is another example that shows that the development of posthuman ethics is directly related to the question of subjectivity. Thus, the development of the rights of novel beings and definition of their moral status require the development of the concept of subjectivity of novel beings and rethinking the Western philosophical conception of subjectivity which originates from the Enlightenment. A similar argument is made by animal activists who claim that animals are also persons (Mitra 2015). Nevertheless, for "hundreds of years there has been a legal wall that separates all nonhuman animals from human beings" (Mitra 2015). Thus, the notion of the subject should include the non-human as well. "This enables recognition that technologies play a fundamentally mediating role in human practices and experiences, and for this reason, it can be argued that moral agency is distributed over both humans" and non-humans (Hanson 2009, 93). However, as it is argued by numerous authors, that moral agency requires consciousness (Himma 2009, 24) and freedom of action (Hanson 2009, 93).

Critical posthumanism reviews the basic tenets of Enlightenment humanism, with the aim of constituting a new ethics, ontology, epistemology, and politics for the posthuman era. Blurred boundaries between the technological and the human, the human and non-human, the natural and artificial, which is a direct consequence of the scientific and technological revolution of the second half of the twentieth century, destabilize traditional concepts of personal identity and social community; ideas about what is inherent in human beings and how they differ from other beings, and above all, the traditional image of the interdependence of body and mind as fundamental strongholds of identity (Zivkovic 2012). This opens the space to create a new ontology-based on hybrid identity instead of the unified, coherent, autonomous, self-conscious, rational being of humanism; for the new teleology, as process and emergence, instead of state; for the new ethics based on new conceptions of fluid identity which overcome sharp binary oppositions and bounded categories (Zivkovic 2012). On the other hand, space is also being created to develop a new epistemology that is not based on fixed categories and irreconcilable binary hierarchies.

According to Carlson, "the posthuman subject is an indeterminate, irreducibly relational, and endlessly adaptive figure whose intelligence and agency are not simply possessed or controlled by the individual or his will, but are distributed throughout complex networks that exceed, even as they constitute, [that] individual", while the subject of modern Western liberalism "is self-governing individual, who exists as such 'by nature' and thus prior to the involvements of social beings and its prosthetic supports" (Carlson 2008, 15). Carlson emphasizes that abandoning the idea of the modern western liberal subjectivity does not mean abandoning humanity—on the contrary, it means broadening the concept of humanity and widening our place in the world (Carlson 2008, 16).

10.3   The Development of Posthuman Ethics from the Postmodern Idea of Subjectivity

A new, posthuman ontology requires a "new, more inclusive form of ethical pluralism" (Wolfe 2010, 137). It opens up the possibility for some new identity politics and clearly shows that posthumanism has its theoretical foundation in the postmodern conceptions of the subject.

The postmodern perspective on ethics is based on deconstruction of modernist notions of subjectivity, justice, and rights (Derrida 1974). Postmodernity advocates the notion of difference, which is perceived as fluid and heterogeneous. It also advocates care and responsibility towards the Other (Lévinas 1969). According to Emmanuel Lévinas, moral considerations are based on social interactions and relationships, not on an ontological basis (Lévinas 1969). Thus, the question of moral and social status transcends definitions and perceptions of the Other (Gunkel 2014).

Critical posthumanism advocates methodology that has its conceptual foundation in the postmodern understanding of the subject, which is viewed as a discursive construction (Zivkovic 2012). In this way, it will open up the possibility for creating some new identity politics based on a fluid and polyphonic concept of identity. This shows that posthumanism is in accordance with postmodernist conceptions of the subject. A new posthumanist ethics and politics require postmodernist understanding of subjectivity, which leaves room for hybrid identities.

Jacques Derrida and Emmanuel Lévinas "in different ways, articulate forms of ethics and justice that move beyond traditional conceptions of subjectivity" (Popke 2003, 299). The posthuman era poses a number of ontological and epistemological challenges which can be resolved by postmodern and poststructural ethics. The

modern ethics was based on the fixed notion of subjectivity "fundamentally constituted through the maintenance of boundaries, both social and spatial" (Popke 2003,
302). Modernist ethics is based on exclusionary categories which stem from the
sharp binary opposition self/other.. The project of modernism clearly separated politics from ethics, right from good, public from private, human from non-human. The
role of postmodern ethics is reflected in the reconciliation of these notions. Postmodern ethics emphasizes the importance of diversity and respect for the Other. It
is based on the conception of personhood in the broadest sense, thus overcoming
the shortcomings of modern liberal theory of citizenship, which assigns rights to an
abstract, rational subject, thus excluding the whole range of individuals and social
groups (Semprini 1997).

According to Emmanuel Lévinas, moral considerations are based on social interactions and relationships, not on ontological basis (Lévinas 1969). Thus, the question
of moral and social status transcends definitions and perceptions of the Other (Gunkel
2018).

Emmanuel Lévinas develops the ethics which transcends the modernist exclusionary idea of subjectivity. According to Lévinas, ethics arises from our responsibility for the other. Lévinas argues that ethics should have primacy over ontology
(Lévinas 1969). He argues that our subjectivity stems from "this pre-ontological
relation to alterity" (Popke 2003, 303). Lévinas states: "My ethical relation of love
for the other stems from the fact that the self cannot survive by itself alone, cannot
find meaning within its own being-in-the-world, within the ontology of sameness..."
(Lévinas and Kearney 1986, 24). Unlike modernist ethics which decide moral consideration "on the basis of some pre-determined ontological criteria or capability",
Levinasian ethics determines moral consideration in light of social interactions and
relationships (Gunkel 2018, 96). Thus, moral consideration is not decided in light of
the question "what the other is in its essence" (Gunkel 2018, 96).

Lévinas's philosophy aims at "transformation of ontology into ethical metaphysics" (Poleshchuk 2010, 2). Lévinas moves away from the question of being
to the question of the Other and revises the European philosophical approach to
subjectivity (Poleshchuk 2010). In his Totality and Infinity, Lévinas emphasizes that
the history of European thought represents a history of discrimination and marginalization of the Other in which alterity was tried to be reduced to the same (Lévinas
1969).

In his essay "The animal that therefore I am (more to follow)", Jacques Derrida
proposes a new ethics that transcends the Western idea of subjectivity based on
criteria of rationality (Derrida 2002). According to Derrida:

> Beyond the edge of the so-called human, beyond it but by no means on a single opposing side,
> rather than "the Animal" or "Animal Life," there is already a heterogeneous multiplicity of the
> living, or more precisely (since to say "the living" is already to say too much or not enough)
> a multiplicity of organizations of relations between living and dead, relations of organization
> or lack of organization among realms that are more and more difficult to dissociate by means
> of the figures of the organic and inorganic, of life and/or death. These relations are at once
> close and abyssal, and they can never be totally objectified. They do not leave room for any
> simple exteriority of one term with respect to another. It follows from that that one will never

have the right to take animals to be the species of a kind that would be named the Animal, or animal in general (Derrida 2002, 399).

Derrida's ethics overcomes binary oppositions: human/non-human, subject/object, natural/artificial, and so forth (Derrida 2001). The purpose of Derrida's deconstruction is transformation of the hierarchical structures which create the metaphysical character of philosophy. Derrida's ethics transcends the discourse based on the power of reason. Derrida argues that the politics of a (fixed) identity, which privileges unity, represents dangerous ethics and politics because of its exclusionary nature (Caputo 1997, 13). The new conception of personhood based on postmodern philosophy and ethics implies overcoming the distinction between the "human" and the "non-human" in legal and political discourse.

10.4   Conclusion

This chapter highlights the relationship between posthumanism and transhumanism. While transhumanism is anchored in the humanistic tradition to which it adds on the transhumanist way of looking at the world, posthumanism represents a break or at least a critical departure from humanism (Furjanic 2020). Yet, as Sorgner points out, transhumanism contains elements that may be relevant to post humanists, just as posthumanism contains elements that may help transhumanists (see: Sorgner 2014). Although it is useful to keep in mind the existence of a certain overlap and mutual influence between posthumanism and transhumanism, as well as the relevance of postmodern ethics for the development of both approaches, they represent different points of view and should be studied separately (Furjanic 2020).

This chapter also argues that the question of the moral status of new beings has often been linked to the question of the ontological status of new beings. Some authors believe that the discussion of the ontological status of new beings should be abandoned, in order to give priority to ethics in light of Lévinas's idea that ethics should take precedence over ontology.

Postmodern theory offers an adequate theoretical framework for the development of posthuman ethics, which is relevant for both posthumanism and transhumanism because it problematizes the binary opposition human/non-human.

According to postmodernist thinkers, biological traits such as gender or race, which are considered as natural and essentialist by modernist thinkers, are constructed by discourse. Postmodern thinkers reject the main concepts of Western metaphysics, such as: subject, identity, truth, reality, and so forth. They argue that that these concepts should not be perceived as fixed, but that they are in need for reinterpretation and deconstruction. Postmodernist authors reject essentialist notion of identity and argue that identity is a dynamic, hybrid, and changeable category. Posthuman ethics requires a postmodern understanding of morality based on the postmodern idea of subjectivity that accepts otherness and overcomes binary hierarchies based on power relations.

Contemporary science and technology are fundamentally changing the self-understanding of human being and life. The question of moral rights and duties is often tied to the question of subjectivity. Proponents of postmodernism advocate ethics which precedes ontology. According to Emmanuel Lévinas, moral considerations are based on social interactions and relationships, not on ontological basis.

References

Balkan, Aral. 2017. Universal declaration of cyborg rights. https://cyborgrights.eu/. Accessed 2 Feb 2020.

Bostrom, Nick. 2003. Transhumanist values. https://www.nickbostrom.com/ethics/values.html. Accessed 28 April 2021.

Butler, Judith. 2001. Contingent foundations: Feminism and the question of "postmodernism". In *Feminism in the study of religion: A reader*, ed. D.M. Juschka. London: Continuum.

Caputo, John D. 1997. *Deconstruction in a nutshell: A conversation with Jacques Derrida.* New York: Fordham University Press.

Carlson, Thomas A. 2008. *The indiscrete image: Infinitude and creation of the human.* Chicago: University of Chicago Press.

Collinson, David. 2006. Rethinking followership: A post-structuralist analiysis of follower identities. *The Leadership Quarterly* 17 (2): 179–189. https://doi.org/10.1016/j.leaqua.2005.12.005.

Cyborg Citizenship. n.d. https://ieet.org/index.php/tpwiki/Cyborg_citizenship. Accessed 5 April 2020.

Denton, Peter H. 2014. The machine question: Critical perspectives on AI, robots, and ethics. *Essays in Philosophy* 15(1): 179–183. https://doi.org/10.7710/1526-0569.1497.

Derrida, Jacques. 1974. *Of grammatology.* Baltimore: Johns Hopkins University Press.

Derrida, Jacques. 1986. Declarations of independence. *New Political Science* 7 (1): 7–15. https://doi.org/10.1080/07393148608429608.

Derrida, Jacques. 2001. *On cosmopolitanism and forgiveness.* London and New York: Routledge.

Derrida, Jacques. 2002. The animal that therefore I am (more to follow), trans. David Wills. *Critical Inquiry* 28(2): 369–418. https://doi.org/10.1086/449046.

Descartes, René. 1993. *Discourse on method and mediation on first philosophy*, trans. D. A. Cress. Indianapolis: Hackett Publishing Co.

Furjanic, Lovro. 2020. Transhumanism—Philosophical foundations. https://hrcak.srce.hr

Gibbons, Michael T. 1991. The ethic of postmodernism. *Political Theory* 19(1): 96–102. https://doi.org/10.1177/0090591791019001006.

Gunkel, David J. 2012. *The machine question: Critical perspectives on AI, robots, and ethics.* Cambridge, MA: MIT Press.

Gunkel, David J. 2018. The other question: Can and should robots have rights? *Ethics and Information Technology* 20 (2): 87–99. https://doi.org/10.1007/s10676-017-9442-4.

Gunkel, David J. 2014. Do machines have rights? Ethics in the age of artificial intelligence. Aurora Online with David Gunkel, interview by Paul Kellogg. http://aurora.icaap.org/index.php/aurora/article/view/92/114. Accessed 15 April 2020.

Hanson, F. Allan. 2009. Beyond the skin bag: On the moral responsibility of extended agencies. *Ethics and Information Technology* 11(1): 91–99. https://doi.org/10.1007/s10676-009-9184-z

Haraway, Donna. 1987. A manifesto for cyborgs: Science, technology, and socialist feminism in the 1980s. *Australian Feminist Studies* 2 (4): 1–42. https://doi.org/10.1080/08164649.1987.9961538.

Haraway, Donna. 2016. *A cyborg manifesto: Science technology and socialist-feminism in the late twentieth century*. Minneapolis: University of Minnesota Press.

Hassan, Ihab. 1977. Prometheus as performer: Toward a posthumanist culture? A university masque in five scenes. *Georgia Review* 31 (4): 830–850.

Hauskeller, Michael. 2016. *Mythologies of transhumanism*. London: Palgrave Macmillan.

Hayles, N. Katherine. 1999. *How we became posthuman: Virtual bodies in cybernetics, literature, and informatics*. Chicago: The University of Chicago Press.

Herbrechter, Stefan. 2021. Before humanity: Or, posthumanism between ancestrality and becoming Inhuman. In *Transhumanism and posthumanism in twenty-first century narrative*, ed. Sonia Baelo-Allué and Mónica. Calvo-Pascual, 20–32. London: Routledge.

Himma, Kenneth E. 2009. Artificial agency, consciousness, and the criteria for moral agency: What properties must an artificial agent have to be a moral agent? *Ethics and Information Technology* 11(1): 19–29. https://doi.org/10.1007/s10676-008-9167-5

Howard, Rhoda E., and Jack Donelly. 1986. Human dignity, human rights, and political regimes. *The American Political Science Review* 80 (3): 801–817. https://doi.org/10.2307/1960539.

Koljenik, Dragana. 2014. Technological utopianism as mainstream: is posthumanism the final construction of man? https://core.ac.uk/download/pdf/197551453.pdf. Accessed 28 April 2021.

Levin, Susan B. 2021. *Posthuman bliss?: The failed promise of transhumanism*. New York: Oxford University Press.

Lévinas, Emmanuel and Richard Kearney. 1986. Dialogue with Emmanuel Lévinas. In *Face to face with Levinas*, ed. R. Cohen. Albany: State University of New York Press.

Lévinas, Emmanuel. 1969. *Totality and infinity: An essay on exteriority*, trans. A. Lingis. Pittsburgh, PA: Duquesne University.

Locke, John. 1836. *An essay concerning human understanding*, vol. I. London: T. Tegg & Son.

Macklin, Ruth. 2003. Dignity is a useless concept. *British Medical Journal* 327: 1419–1420. https://doi.org/10.1136/bmj.327.7429.1419

Manzocco, Roberto. 2019. *Transhumanism—engineering the human condition: History, philosophy and current status*. Chichester, UK: Springer in association with Praxis Publishing.

Mitra, Maureen Nandini. 2015. Animals are persons, too. https://www.earthisland.org/journal/index.php/magazine/entry/animals_are_persons_too/. Accessed 2 May 2021.

More, Max and Natasha Vita-More. 2013. Roots and core themes. In *The transhumanist reader*, ed. Max More and Natasha Vita-More, 1–2. Chichester, West Sussex, UK: Wiley-Blackwell.

Morgan, Wayne. 2000. Queering international human rights law. In *Sexuality in the legal arena*, ed. Carl Stychin and Didi Herman, 208–225. London: Athlone Press.

Nowak, Manfred. 2001. Foreward. In *Humiliation, degradation, dehumanization: Human dignity violated*, ed. Paulus Kaufmann et al., v–vi. Dordrecht: Springer.

Pilsch, Andrew. 2017. *Transhumanism: Evolutionary futurism and the human technologies of utopia*. Minneapolis: University of Minnesota Press.

Poleshchuk, Irina. 2010. Heidegger and Levinas: Metaphysics, ontology and the horizon of the other. *Indo-Pacific Journal of Phenomenology* 10 (2): 1–10. https://doi.org/10.2989/IPJP.2010.10.2.4.1085.

Popke, E. Jeffrey. 2003. Poststructuralist ethis: Subjectivity, responsibility, and the space of community. *Progress in Human Geography* 27 (3): 298–316. https://doi.org/10.1191/0309132503ph429oa.

Pruchnic, Jeff. 2013. *Rhetoric and ethics in the cybernetic age: The transhuman condition*. London: Routledge.

Philosophy Question. 2019. What is known as humanism?, https://philosophy-question.com/library/lecture/read/91543-what-is-known-as-humanism. Accessed 3 May 2021.

Roden, David. 2010. Deconstruction and excision in philosophical posthumanism. *The Journal of Evolution & Technology* 21 (1): 27–36.

Roden, David. 2014. Brandom and posthuman agency an anti-normativist response to bounded posthumanism. https://enemyindustry.wordpress.com/2014/08/23/robert-brandom-and-posthumanism/. Accessed 5 May 2020.

Schroeder, Doris. 2012. Human rights and human dignity: An appeal to separate the conjoined twins. *Ethical Theory Moral Practice* 15: 323–335. https://doi.org/10.1007/s10677-011-9326-3.

Searle, John. 1984. *Minds, brain and science*. Cambridge, MA: Harvard University Press.

Searle, John. 2002. Can computers think? In *Philosophy of mind: Classical and contemporary readings*, ed. David Chalmers, 669–675. Oxford: Oxford University Press.

Searle, John. 1980. Minds, brains, and programs. *Behavioral and Brain Sciences* 3(3): 417–424. https://doi.org/10.1017/s0140525x00005756

Semprini, Andréa. 1997. *Le multiculturalisme*. Paris: PUF.

Sorgner, Stefan Lorenz. 2014. Pedigrees. In *Post- and transhumanism: An introduction*, ed. by Robert Ranisch and Stefan Lorenz Sorgner. Frankfurt am Main: Peter Lang.

St. Pierre, Elizabeth A. 2000. Poststructural feminism in education: An overview. *Qualitative Studies in Education* 13(5): 477–515. https://doi.org/10.1080/09518390050156422

Thomas, Alexander. 2017. Super-intelligence and eternal life: Transhumanism's faithful follow it blindly into a future for the elite. The Conversation. https://theconversation.com/super-int elligence-and-eternal-life-transhumanisms-faithful-follow-it-blindly-into-a-future-for-the-elite-78538. Accessed 10 Aug 2020.

United Nations. 1948. Universal declaration of human rights. https://www.un.org/en/udhrbook/pdf/udhr_booklet_en_web.pdf. Accessed 3 Feb 2020.

Wolfe, Cary. 2010. *What is posthumanism?* Minneapolis: University of Minnesota Press.

Zivkovic, Milica. 2012. To a new world of gods and monsters: Posthumanizam, teorijsko-kritički pravac za novi milenijum. In *Nauka i savremeni univerzitet*, 2, 34–46. Nis: Filozofski fakultet.

Sanja Ivic is a Research Fellow at the Institute for European Studies, Serbia. She completed her Postdoctoral research at the University of Paris 10, France. She is a member of the Editorial Boards of three peer-reviewed international journals: International Law Research (Canada), American International Journal of Contemporary Research (USA), and Journal of Law and Conflict Resolution (Africa). She also cooperates with various international scientific institutes and teams. She was a member of the Steering group of the project "Pluralism, Inclusion, Citizenship" (UK) and she is currently a Board Member of the International Society for Philosophers (UK). Her publications include books and articles on various subjects in the fields of philosophy, political science, and European studies. Her book European Identity and Citizenship: Between Modernity and Postmodernity is published by Palgrave Macmillan/Springer in 2016. Her book Paul Ricoeur's Idea of Reference: The Truth as Non-Reference is published by Brill in 2018, and her book EU Citizenship: Towards a Postmodern Conception of Citizenship? is published by Vernon Press in 2019. She is currently working on a book The Concept of European Values: Creating a New Narrative for Europe that is going to be published by Lexington Books in 2022.

Chapter 11
Can Posthumanism Be Post-sexist?

Sonia Reverter

Abstract What if any limitations should biohacking have in relation to sexualised bodies? Could posthumanism offer a liberating decoding of sexual identity? Will the socially constructed gender binary disappear? Will this allow us to rid ourselves of gender inequality? Is this the path we may envisage from transhumanist discourses and current biohacking practices? This certainly does not seem so. Most discourses and biohacking practices lock us into a hypersexualized world where bodies prevent us from exiting the binary coding of genders into masculine and feminine (currently hacked to appear as "super masculine" and "super feminine"). Donna Haraway was the first to conceive a post-human world that could break the inequality of the binary system in her theory of the cyborg. She proposed the cyborg as "a creature in a post-gender world." However, this vision has not been fulfilled nor is it widespread. As Haraway herself warned, the cyborg may end up falling into the hands of patriarchal capitalism and militarism. In this chapter, I propose an understanding of posthumanism as a possibility for a new materialism that allows us to transcend gender inequalities. To this end, I follow up on the reflection of Haraway and Braidotti to explore a trans social model that may be transgender in a fashion in the way it completely challenges today's binarism. This feminist posthumanism intends to and is able to transcend the binary sex difference as an element of social order. The current potential of biohacking techniques could make sex differences no longer the fundamental differential element in the technologies of power over bodies and lives. In this regard, I propose a critical reflection on ethical commitments, which democratic public institutions should begin to lead so that the posthuman reality may serve justice better than the human one.

S. Reverter (✉)
Castelló de la Plana, Spain
e-mail: reverter@uji.es

E. Tumilty and M. Battle-Fisher (eds.), *Transhumanism: Entering an Era of Bodyhacking and Radical Human Modification*, The International Library of Bioethics 100, https://doi.org/10.1007/978-3-031-14328-1_11

189

11.1 Introduction

I find myself writing this text at a time full of uncertainties and big changes, both on a personal and a planetary level. The pent-up fear and concealed puzzlement during the mandatory lockdowns in response to COVID19 have placed me into a state of mind reigned by perplexity and an inability to concentrate. I had been reading for some time about the possibilities of posthumanism to overcome sex/gender inequalities. Decades ago, one of my favourite philosophers, Monique Wittig, wrote a book, which begins like this: "The perenniality of the sexes and the perenniality of slaves and masters proceed from the same belief, and, as there are no slaves without masters, there are no women without men" (Wittig 1992: 2).

I had begun to conceive a world towards which I thought we were ideally heading where the sex/gender[1] category would no longer serve as a dealer of roles and privileges. Thus, a more egalitarian world precisely because it deactivates the category that structures identities and life experiences through stereotypes. A better world as an ungendered world.

Suddenly, the family time spent within a global pandemic and a lockdown put me in a new way face-to-face with the question of how to live my life day-to-day in a present that had nothing to do with my reality just one day before. This new situation presented a new side of the vulnerability of humanity and impels us, more than ever, to undertake the philosophical task of reflecting on how to conceive a world that can respond to humanity in danger. I understand this task as a task of critical thinking; to paraphrase Foucault (2001) on the meaning of critical, it is about making hard that which looks easy. My reflection in this chapter is to think about this present and future in a critical way so that we may overcome one aspect at the core of humanity's vulnerability: the inequality that we keep building in relation to the sex/gender category.

My point of departure is understanding that humanism is not enough. It has been useful and has helped with significant human improvements. However, it is a spent theoretical framework for the global issues that humanity is facing today. It is a discussion that fails to address the question of inequality between women and men meaningfully. In failing to address this, humanism fails to be relevant to our present moment, when we know, as shown by UN reports (UNWomen 2020a, b; UNDP 2012, 2020), that gender inequality lies at the heart of many of the greatest global issues. That is why my reflection here focuses on the possibilities of a post-sexist posthumanism: the possibility of an equal world on the basis of gender and sex.

Given the dystopic present we have, we can see how posthumanism and transhumanism[2] that have been discussed for decades and which were getting more attention

---

[1] I will generally use '*sex/gender*' as two terms referring to a biosocial categorization. Therefore, they cannot be separated when mentioned, discussed or experienced. We can only separate them for abstract purposes.

[2] The definition of the terms "posthumanism" and "transhumanism" is in itself a matter of debate. We can find texts that clarify any differences and diversities as far as opinions on this question (Ferrando 2013). Both terms, trans- as well as posthumanism share the notion of technogenesis

in academic departments, conferences, and journals, along with biohacking and artificial intelligence development is being reflected on the news. Some of the images that we are seeing from some countries that are using artificial intelligence and big data to curb the spread of the virus are telling us about a new reality. A reality in which technology somewhat independently works alongside human beings in dealing with crisis. I wonder what we can do, where to look, to transform these possibilities and push towards an ideal for humanity that should not be dismissed as unachievable even if it is currently utopian. It is important from my point of departure to acknowledge the essential work of critical thinking. The question of a post-sexist world is mandatory if equality remains as a regulatory ideal. The posthumanism that I advocate here is, in this way, critical of humanism as an individualistic and human-centred framework, but still connected to an egalitarian aspiration. As Hayles (2011: 226) states: "Imagining the future is never a politically innocent or ethically neutral act. To arrive at the future we want, we must first be able to imagine it as fully as we can, including all the contexts in which its consequences will play out".

I think we need to organize life differently, and to do this the human aspect should not be our focus, when we direct our attention towards the world we are creating. We should look at the world (and create it) not focusing on the human but quite influenced by interests that are a threat to the human race and all the other life on the planet.

The planetary deficiencies and problems we have been experiencing for some time have become more critical precisely due to a global pandemic that has arrived overnight. In a way, this current time is so crucial that some concerned thinkers are already talking about a crisis of civilization. Some years before the pandemic, the prestigious archaeologist and paleontologist Eudald Carbonell already announced the systemic change we are facing. This change does not entail an evolution of our current world—Carbonell says that it is a "different world" (2018). What does he mean by this? The difference is subtle and may be useful to reflect on in posthumanism. The definition and history of the questioning what it means to be "human" is difficult to entangle as its history dates as far back as the very *Epic of Gilgamesh*, 3,700 years ago, as we are told by Bostrom (2005), one of the most renowned transhumanists. In this

---

(meaning humans in coevolution with technologies). And both concepts question humanism and reconsider what it means to be human. The debate over the terms is on how to interpret the meaning of a "post" human state. This debate gets complicated if we notice that both terms are understood in a different way being North America or Europe, as Ranich and Sorgner (2014) discussed. Since these terminological questions are beyond the scope of this chapter I will follow the definition of both terms given in Ranich and Sorgner edited book (2014), where transhumanism intends to create a bioliberal future, or as Sorgner (2014) calls it "a carbon-based transhumanism" heading to bodily improvement. As Hughes explains in his chapter in that book (2014: 145) transhumanism wants "to free themselves from the constraints of nature and fulfill their own concepts of the good life". The reflection on transcending the dichotomy of the sexes within feminist traditions has been addressed mainly not from a technological and biological perspective, but from a social criticism that challenges hierarchical social construct assumptions and deeply questions categories. I will use the term "posthumanism" since it connects better with this philosophical background I want to stress. Furthermore, it is related to an essential aspect of my reflection, the feminist critique to patriarchal humanism (Nayar 2014).

sense, everything human is in constant evolution and it is, we could say, transcended. What is different now? Why the evolution is now more than an evolution? What is new now if we are always evolving as humans? In the *Transhuman Manifesto,* Vita-More tells us:

> Humanity needs a change—a new outlook that helps us become more humane. This something new is transhumanism—a worldview that seeks a quality of life that brings about perpetual progress, self-transformation, practical optimism, visionary solutions, and critical thinking—the transhuman". (Vita-More 2020, v 4)

These small lines do not clarify what the quality of life that transhumanism (and post-humanism) allegedly aims for. In a way, this could have even been said in other times in history, as Bostrom points out. That is why I find Carbonell's proposal important: a different world with another paradigm. This paradigm is what Carbonell calls "species consciousness."[3] In other words, Carbonell affirms, as an expert in prehistory and the humanization process of the *homo sapiens,* that today we may avoid the parsimony of natural selection and make cultural and social selection work for the benefit of what he calls "critical species consciousness." This consciousness must be critical meaning that we recognize ourselves as an intelligent species and can socialize such intelligence to live on the planet in balance—as a human community, a balance with other species and the ecological system that sustains us. This species consciousness, critical and socializing, will lead us to our becoming posthuman. Biotechnology will prove essential in this different world since it will have to generate (and protect) the necessary diversity for the degree of sociability required to live in balance. A biological and cultural diversity implies and creates more equality, for it is the only way to generate species consciousness. In various interviews and videos, Carbonell brilliantly exposes how uniformization and hierarchy are enemies of that posthumanist future of species consciousness.

This conceptual framework allows me to place diversity and equality as central to a posthumanism and a posthumanistic world. Many otherproposals considered posthumanist do not gather, in my opinion, the necessary conditions to call themselves such, because they do not help us see a world free from inequality amidst the ways in which life relates to itself in all its diversity.

I turn away from transhumanist (and posthumanist) stances that attempt to recreate fantasies as to how to transcend everything human, such as the body and the soul, in order to reach immortality or to achieve bodies with unlimited powers. This way of conceiving posthumanism, criticized by such authors as Katherine Hayles (1999), is not useful for a kind of posthumanism that wants to overcome concepts that have colonized the interpretation of body and nature.

> If my nightmare is a culture inhabited by posthumans who regard their bodies as fashion accessories rather than the ground of being, my dream is a version of the posthuman that embraces the possibilities of information technologies without being seduced by fantasies of

---

[3] Decades ago, Carbonell began to air his concern about the future of the human race. He used the concept of "critical species consciousness" in his extended work as a disseminator during talks, conferences and interviews. He explained this in writing in his (2018) book and a recent (2019) book (in co-authorship).

unlimited power and disembodied immortality, that recognizes and celebrates finitude as a condition of human being, and that understands human life is embedded in a material world of great complexity, one on which we depend for our continued survival (Hayles 1999: 5).

Thus, I understand posthumanism as linked to a political conception of envisioning attitudes, of materializing and becoming a community, which transcends the limits of a biopolitical system as a knowledge producer that causes and maintains growing inequalities; and, specifically, oppression through a sex/gender code in its masculine/feminine binarism. As I see it, this requires giving up the biopolitical system that creates and codifies identities within a sex/gender binary system. This entails a revision of one of the concepts that are used in connection with posthumanist and transhumanist proposals: the biohacking concept. All this cannot be done without an epistemic rupture that feminist theory has for some time deemed necessary to progress within the material equality between women and men (Haraway 1985; Fox Keller 2010; Segato 2003).

Donna Haraway[4] herself (2006)—a pioneer when it comes to showing the possibilities of the epistemological ruptures that sustains this binarism—has criticized the ideas of a posthumanism that caricatures its concept and philosophy of the cyborg in a new "techno-masculinism". Braidotti (2013) has also reflected on the new role of the posthuman humanities from feminist theory. Among other challenges, we should respond to the question of how posthumanism may develop new ways of engendering the subjectivity.

In her reflection on bio-enhancement and gender, Reilly (2018) warns us about this risk. It is for this reason that the very definition of posthumanism is a complex task though essential to understand in what way it may or may not transcend the sex/gender categorization of humans. Reilly focuses on specific aspects that challenge the binary system of assigning a sex and gender identity such as intersex or trans people. This is a necessary path when presenting the debate on biohacking gender with regard to posthumanism. However, I will not deal with its specific aspects in this chapter. The realities of transgender and intersexuality, with all their biological and social diversity, may help to provide ways of liberation for other bodies (Reilly 2018: 276). However, the framework from which I am reflecting here does not allow me to deal with medical questions or questions of "improvement" (or bio-enhancement, the term used by Reilly to entitle her paper). Overall, I will focus on the rupture of a normality that imposes a bimodal way to understand bodies.

I position myself within a theoretical-practical context to conceive of posthumanism as it is proposed by Cecilia Aasberg and which is summed up in the following quote:

In the posthuman context of the Anthropocene, I suggest and point to postdisciplinary humanities research and theory–practices that pay careful attention to the feminist theoretical work on our equally postnatural condition as an experimental remedy (Aasberg 2017: 185).

---

[4] Haraway no longer uses the term "posthumanism" today as she did in her first works. However, I interpret Haraway's thought in a way that the ideas she has developed from her first essays up to the present day form a solid and coherent core for the proposal of the feminist and post-sexist posthuman that I am defending here.

This text by Aasberg is especially interesting in order to understand what the author rightly calls "feminist post-humanities." In it, Aasberg demands the usage of the long hermeneutic tradition found in feminist theory "to tackle our posthuman, and equally postnatural, condition as an open-ended biosocial event" (2017: 186). This tradition places my interest in Donna Haraway as the first example of feminist reflection on posthumanism. In her work, we find a new conception about the materiality that allows us to understand each other as biosocial beings and, hence, as biopolitical beings. This point allows me to collect an ethical–political proposal to conceive of posthumanism. A posthumanism that is ethical and political because it is proposed as a just way of living whose ecological milieu and niche—and not just its horizon and ideal—shall be equality and diversity. Or, in Carbonell's words, the critical species consciousness, as I have already mentioned.

11.2   Biohacking as a Biopolitical Practice

For my task of delving into the possibilities that posthumanism and biohacking possess to decode the system of human inequality with regard to the sex/gender binary identity, I will refer, as a starting point, to a straightforward definition of "biohacking" provided by Vita-More (1983): "Humans breaking away from their biological limitations".

My interest is to reflect on how, by breaking the biological limitations that we have, we can form a posthumanity where inequality does not take place, including sex/gender-based inequality. Therefore, posthumanism and its implied relationship with biohacking allow me to understand the posthumanist proposal as a proposal to overcome inequalities linked to the values of democracy. Posthumanism as a solution to the weariness of humanism.

The goal of biohacking, in the first place, is to improve the capacities of the human body by manipulating it with the usage of supplements and/or technology. Secondly, to put the knowledge, techniques, and technology required to do this within the reach of everyone. In other words, to make available for everyone that which many times only occurs in laboratories owned by large corporations reserved for the elite few. Democratizing cutting-edge "science" while keeping it safe. As a feminist, I am quite drawn to the power of resistance and social mobilization brought by biohacking action. In other words, I understand that the improvement of human capacities by manipulating the body helps us create a more democratic society not just in the usage of science, but also in the political, social, and cultural goals that this manipulation of bodies has. I am interested in finding out whether biohacking practices are sufficiently disruptive to generate a movement that would deliver that equality to the human species.

I focus specifically on how biohacking may be useful for us to overcome inequalities based on sex/gender. We should consider this a great challenge as this is one of the most universal and most persistent inequalities. Furthermore, the thesis of sex/gender-based inequality usually relies on the support of scientific explanations, which hold sex differences as "natural" causes that partly explain and justify inequality by being validated by scientific authorities. The so-called "neurosexism" (Fine 2010), found in some neuroscientific publications, would be an example of this (Fausto-Sterling 2000; Rippon 2019; Joel et al. 2015). Scientific networks that are critical of neurosexism, such as *Neurogenderings Network*, help us confirm the need to dismantle the prejudices that science itself maintains concerning sex difference (Reverter-Bañón 2017b, 2019).

In some ways biohacking itself can be seen as sexist as it can refer to DIYbio techniques related to diets and body sculpture or to the self-management of reproductive capacities for women. I refer, explicitly, to the common examples of how the managerial discourses of the market and consumption if allied with those holding power, such as policy makers and healthcare professionals, impose a mandate on bodies and health that negates the information and autonomy needed to decide about one's own body, as it is denounced by the reproductive justice advocates. In both cases, body transformation and reproductive interventions, the main message that guides the majority of biohacking practices is to adapt and "improve" women's bodies to the demands of the neoliberal patriarchy. As regards this, we can link the biohacking culture mostly to an aspirational individual culture of wanting to be a better different self, according to the rules that have been pre-established in a sexist hypersexualized world.

I want to stress the fact that many current biohacking practices not only fail to take us to the posthumanism that I defend, but that they even fail to be disruptive in their own sense by not going beyond the current rules in a biopolitical sense. At this point, I think we begin to see that the first definition of biohacking that we obtained from Vita More ("Humans breaking away from their biological limitations") falls short. In what way? In the way in which we know today that the biological limitations are not only biological but also political. Feminism is quite aware that there is no way to separate the body and its materialization from the political. Furthermore, the system of oppression against women, transgender, and non-binary people—the patriarchy—is a scheme that intertwines the political with the biological. As some feminists have been pointing out for decades (Butler 1990; Wittig 1992), how could we detach one from the other? Sex difference (the one that the patriarchy employs as its basis to generate inequality) is traversed by both, the biological (sex) and the political (gender). In reality, the relationship between sex and gender is, as pointed out by Grosz (1994), a Möbius strip in full motion—it is difficult, if not impossible, to know whether you are entering one or the other. Thus, we categorize the biological sex according to a masculine/feminine bimodal model in tune with a social construct of man/woman. This is the so-called 'sex/gender system' where inequality between women and men is maintained.

We must appeal to a definition of biohacking as a biopolitical action. In other words, in its biological sense (as an improvement of the capacities of the human body) as well as in its political sense (as a movement to put all the knowledge, techniques, and technology required for such improvements at the reach of everyone so that they can serve egalitarian and democratic causes and disrupt current power). There is no way to transcend the limits of the ruling biopolitical system concerning the sex/gender bimodal categorization with just a conception of biohacking that is solely linked to the biological. That is why we need to understand the potential of biohacking within the framework of a conception of critical biopolitics.[5]

In order to extend this argumentation, we shall refer to Foucault's conception of knowledge. Foucault (2001) stated during his lectures on the concept of governmentality at the *College de France* in the 1970s—that knowledge is not so interesting as an ideology or a theory, but as practices produced by the relationships of power. The separation of power and knowledge is a myth that Nietzsche helps us undo (Gordon 2001). I am interested in underlining that the modern political rationale is expressed through that conception of politics as knowledge practices that entail a political arithmetic, which Foucault relates to statistics (2009), but today it would also be logarithmic and related to the Big Data and AI, as D'Ignazio and Klein assert (2020). For our own interest, the development of these ideas places us in the field of biopolitics. Biohacking as biopolitics enables us to find consensus and carry out emancipatory (knowledge) practices.

The ruptures, discontinuities, and transformations that we can produce with biohacking techniques will have little social repercussion unless we manage them in a way that they can have an impact on the biopolitical regime of power and knowledge practices. Feminism is a niche of knowledge that is especially committed to this idea as it knows, through its own history, that power and knowledge practices are the most resilient to transformation. What kinds of issues may we find today if we finally decide to get rid of a stereotyped and dichotomic view of the genders. Why does such a technological possibility for human improvement such as biohacking not serve liberating practices instead of serving those that are renormativizing and patriarchal? Why is biohacking not used to explore ways to overcome biosocial categorizations that imply inequality and exclusion?

It is important to remember that the goal of every feminism is to transform society in order to make it egalitarian. To this end, the organizational skills of feminism have been aimed at proposing, launching, and demanding changes. And all this has been done with a clear social objective as well as one towards human improvement.

Let us take, as an example, the concept of 'performativity' which Butler began to use in the 1990s (Butler 1990). Performativity is understood as the capacity to act from identity itself. I am a woman or a man because I carry out the action of being a woman or a man. These actions and behaviors form 'gender', feminine and masculine. For these two genders, there is a series of established rules that are introduced into our understanding of being a woman or a man. As these actions are repeated and performed, we socially become women or men based on the assignment that is established according to the binary parameters with which we interpret the

---

[5] The concept of "critical biopolitics" has been introduced by Makarychev and Yatsyk (2020), in analogy with other critical streams of contemporary political thought. They define it in this way: "Critical biopolitics thus denotes a type of academic discourse that, first defies and debunks the binary type of thinking as simplistic and reductionist. Instead of binaries, it conceptualizes the social world as an endless series of complex chains of distinction, correlations, and correspondences. Critical biopolitics welcomes post-foundational and rhizomatic variability of forms of life and modes of politics, as well as intersections between them" (Makarychev and Yatsyk 2020: 24),

body from birth. In other words, what we learn from the concept of performativity is that we form gender by repeating rules that are legitimized from power and knowledge practices, and this massive and collective repetition strengthens these rules which eventually appear as almost "natural".

The question on which we are focusing now is to point out how the great conceptual wealth that this theory of gender performativity may have must be linked to a social project, to a collective action, and, in short, to a political goal, in order to make the best of its potentialities (as Butler herself does in 2015).

It is collective action through the rules which form and legitimize identities, so that only collective action may transform societies. And this is an important lesson when conceiving notions that imply, at their very core, a sense of change, as is the case with the concept of biohacking. With the concept of performativity learned in feminist theory, I would like to point out the fact that only through collective dissident practices—irrespective of whether they were started by collectives or individuals—can we expect significant social changes (Butler 2015). Can biohacking practices propose a nonsexist view of the body? Yes, they can. And there are some inspiring examples of this. I would like to quote Olivares (2014) as one of these examples. Using the "transbecoming" concept created by Eva Hayward (2010), she carries out a project entitled "10,000 Generations Later." She describes it as follows (2014: 296):

> A speculative project of the flesh, a subdermal time capsule, and archive inspired by Octavia Butler's multispecies storytelling and in particular, the co-evolutionary potential she projects onto corporeal archives. My intention with this project is to construct a subdermal archive composed of the matter that has influenced the evolution of my consciousness.

This project has a post-essentialist aim, as the author affirms, and a miscegenational one, beyond the human species. Thus, in my opinion, Olivares claims a transforming liberating action that must necessarily break the division beyond the sexes/genders in the same way that the author who inspires her, Octavia Butler, does in her science fiction works. However, the action itself—as an archive of its consciousness—may not become politically relevant at a social level. In other words, it may not bring about social change. Self-ethnography is indeed a way to mix embodiment and politics, as Stryker (2005) and Hayward (2010) also maintain. However, that individual self-ethnographic proposal must have a collective interpretation and must be replicated (iterated) in time (Butler 1990) to transform the collective categories that shape and govern our reality. Olivares brings up a very interesting point which I want to highlight, and that is, the need to deconstruct the hegemonic archives, to hack them as a necessary political action as Foucault himself proposed. Olivares' proposal (2014)—which I am using as an example—contains the elements required to arrive at that post-sexist world that I yearn for, though they do not suffice in and of themselves.

They do not suffice because for a post-sexist transformation we need social performativity to intercede in the process of creating, defining, and interpreting the materiality (Reverter-Bañón 2017a). There is no way to change our current world into a non-sexist egalitarian one if there is no collective political project or if it is not built performatively. The biohacking proposals that we normally find are not transforming and have no intention to transform in a sense of 'political emancipation,' to use the old terminology that the Frankfurt School already coined back in the 1940s.

The presence of a body and the performative action that may take place within an individual body through biohacking techniques—such as those that are commonly advertised with a promise of human improvement—do not have enough political strength to achieve social change. Thus, from a depoliticized, individualized, and neoliberal view of 'change' and 'transformation,' biohacking ends up looking too much like other perpetuation techniques of gender stereotypes and roles.

From what I am saying this irremediably implies that the biopolitical conditions that enable the development of life (not just human life) must be taken seriously. The governance of the bodies, as Foucault termed it, has, within the performative concept and practice of gender, one of the most important political technologies to determine what we are and to what we may aspire. Thus, I am trying to understand, within this framework of what we call "critical biopolitics", the capacity of biohacking practices to become an ally of a posthumanist proposal that can transcend the representations and figurations of human life in a gendered and unequal way.

11.3   Biohacking as a (Feminist) Social Movement

Thus, I depart from a conception of biohacking that is linked to a way of understanding the governance of the bodies and of life. For this reason, the questions I am interested in are: Can we propose, via biohacking practices, a social change that would allow us to talk about a post-sexist and post-gender, egalitarian and, therefore, posthuman world? We could, but they need to transcend the individual context. Every dissident practice—as it must be to overcome the ruling sexism—needs to be collective or to become collective. It may only become transforming or emancipating if it is collective. Only if a new practice is carried out as a collective movement, may it transform the structures and epistemes that create inequality. Thus, I understand biohacking—in its connection with a biopolitical framework—to go hand in hand with a posthumanist egalitarian commitment. The work of Sánchez Barba (2014: 2) is quite remarkable as it is one of the few that analyzes biohacking as a social movement: "Studying DIYbio as a social movement can generate insight into the dynamics of the movement in terms of how it brings about social change".

Thus, for an egalitarian posthumanism, the contribution of biohacking must not only have a collective motivation, but also a counter-hegemonic one; in a sense of going against the patriarchal hegemony that defines the sex/gender stereotypes that create inequalities. The hegemonic patriarchal order, as Segato (2003) tells us, is so because it is "the air that we breathe." That is, patriarchy, for being hegemonic, has the power to legitimize and confer meaning to identities, to the world, and to knowledge. It is the framework through which we confer meaning to all, from our lives to the very meaning of words. A counter-hegemonic undertaking requires a lot of awareness, attention, and coordinated political action to bring patriarchy down. It must be able to propose techniques and ideas that politicize practices with bodies in order to question the predominant binary categorization of gender and the sex/gender system that sustains inequality. The xenofeminism proposal—as abridged in Laboria Cuboniks' Manifesto (2015) and developed by Hester (2018)—and the present proposal converge at significant points, especially when the Manifesto (Laboria Cuboniks 2015: 2) raises such questions as:

> Why is there so little explicit, organized effort to repurpose technologies for progressive gender political ends? XF seeks to strategically deploy existing technologies to re-engineer the world (…) Technoscientific innovation must be linked to a collective theoretical and political thinking in which women, queers, and the gender non-conforming play an unparalleled role.

Thus, the goal is to create a world where the identity logic that creates constant subordinateness and oppression is terminated. However, in order to do this, we must transform the epistemological question that wonders about the self. Nowadays, there is a great dispute over "what is being a woman" as a result of the ongoing debate that many countries are having on the rights of transgender people and how to fit them into the law as legal subjects. I am not going to get here into the long and complicated history of this question which Stryker herself (2005, 2017)—quoted above—analyzes in her work on trans reality. What I want to point out is that, at this moment when we seem to need to solve such an ontological question in forums and social media, there is still an epistemological stagnation. This stagnation refers to the repetitive machinery of knowledge forms and practices that have, as their only orientation, the creation of fixed categories, thus producing an "epistemology of ignorance" as Tuana (2006) calls it, or a repetition of biases legitimized by science, as neurofeminist critiques clarify for us (Fine 2010; Rippon 2019).

This machinery evinces how we still conceive knowledge as independent of practices and contexts inked to the powers. As Foucault (1969, 2001) warned repeatedly, we still think about knowledge as being self-referential and independent of the social and historical context. However, as Foucault tells us, that which we call 'knowledge' responds to an 'epistemic structure.' This structure is formed within a "group of relationships that exist in a certain period among the different sciences" or "different discourses." This is the ground on which we build the ideas of an historical period.

The patriarchal epistemology that focuses on establishing the meaning of what a woman is through creation and repetition generates an ontology that is incapable of rendering an account of the details of human life itself and its rich diversity. This restrictive ontology would be the wrong path to arrive at the posthumanism that I am defending here. As regards some issues and occasions, we may use, as feminists, a certain "strategic essentialism"—as Spivak (1993) called it—to summon "women," so that we can come together in the struggle, as is the case for gender violence. However, that essentialism refers, in turn, to a regime of knowledge creation, of epistemic structure, that turns women into inferior beings that are to be oppressed. As an identity category, sex serves as ontological substratum to build an epistemology that shapes a dual construct—masculine and feminine. According to this, it is not just a question of being and being in this world, but also a question of understanding, interpreting, and organizing life in it. As several feminist theories have repeatedly warned, this ontological substratum naturalized in the binomial interpretation of gender—with its epistemological consequences—enroots and somehow prevents the deconstruction of the subjects in inequality. For this reason, it is the epistemological system, structure, and regime that govern the ontological inequality anchored in the sex/gender identity.

A posthumanism that transcends the sex/gender mark for its collective organization requires new ways to understand the materialization processes. With her concept of "agential realism," Karen Barad refers to this when she talks about "how matter comes to matter." Materialization comes about through performative collective actions. She describes how essential it is to understand this for a posthumanist proposal:

> The separation of epistemology from ontology is a reverberation of a metaphysics that assumes an inherent difference between human and nonhuman, subject and object, mind and body, matter and discourse. Onto-epistem-ology—the study of practices of knowing in being—is probably a better way to think about the kind of understandings that are needed to come to terms with how specific intra-actions matter (Barad 2003: 829).

In order to contribute to the necessary rupture with this epistemological regime, biohacking must be able to establish itself in collective practices—as I said—and go beyond the separation between biology and culture in a feminist way and with a liberating intention. Every biohacking practice has this ultimate intention, to break down the wall between biology and culture. However, the idea is to do this in a way that liberates the oppressed subjects. Otherwise, breaking down this wall could validate dictatorial, oppressive, and dystopic ideologies and regimes.

The proposal to overcome the barriers of nature and culture with an emancipating purpose concerning sex/gender was already proposed by Donna Haraway back in 1985 with her *A Cyborg Manifesto*. In this text and other works collected in her 1991 book (Haraway 1991), we find the core idea, which has served as a guide to many posthumanist proposals. What she suggested more than three decades ago is the need to conceive a world that could put an end to the inequalities of the gender binary system. To this end, we had to get rid of the biology/culture dualism, but this had to be done carefully so that it would not end up in the hands of patriarchal militarist capitalism. Since the time she proposed these theses back in 1985, Haraway has continued to elaborate ways to dismantle the identity ontology of oppression with profound epistemological changes. In her last books (2007, 2016), Haraway maintains the idea of life as a "becoming-with,[6]" in multispecies accounts and practices, at permanent risk and precarity. These elements are in themselves proposed as enablers of a post-gender world.

I am interested in Malatino's reading (2017) of Haraway when she points out an aspect that usually goes unnoticed when interpreting the cyborg proposal. It is about the need to fight the violence that has formed our colonization, materialities, and accounts. I find this ethical and political aspect in Malatino quite interesting when it comes to an egalitarian posthumanist proposal. The world's material modification to reach this equality is a challenge for the stereotyped vision of biohacking practices

---

[6] This concept is similar in its aims to the concept of "social performativity" explained above.

that oftentimes appear in the media. Normally, the examples that go around most frequently, even those that talk about 'biohacking for females, recreate the bodies in a binary way, maintaining the sex/gender differences without even considering the possibility of hacking the body in one of its most determining categories in our human lives such as that of sex/gender. These recreations constitute, as Malatino tells us, examples of recolonization of sexual difference. Some of the biohacking examples that we may find on the Internet show us a hypersexualized and sexist world, a materially unequal one though equally degraded and impoverished for men and women. This way of showing the possibilities to change the body, to hack it, without even questioning such binarism, form a strategy that is recolonizing in a sense that they reaffirm the binary view of the bodies that patriarchy has interpreted as the only natural one. Patriarchy colonizes the body of women when it interprets them and occupies them as a territory that is owned by males (Segato 2016). This leads to the creation of a colonized body, identity, and sexuality (feminine woman, heterosexual) at the service of a colonizing body, identity, and sexuality (masculine, man, heterosexual). The message that may be sent is: "It is so natural that it is even presented as uncodifiable." It is like a limit that we cannot trespass, not even by hacking it. If there is no code, it cannot be hacked.

In relation to this, Malatino gives us an account of the experiences of trans people as biohacking experiences that may help us understand the negative aspects of this colonization and its violence. If we refer to Preciado's "pharmacopornographic" accounts told in the first person, along with the experiences of people with hormones, we can see the dual reality presented to us. On one hand, that of individuals who easily mold and transform their bodies according to idealized visions of "self-surpassing human." On the other hand, we find individuals who have great difficulties accessing medical, technical, pharmaceutical, or surgical procedures that allow them to meet the "gender-confirming" social demand. Individuals are pressed to remain in a regime of biopolitical recognition that only allows the feminine/masculine duality. We are demanded to follow the biopolitical rule of the epistemic regime, even during procedures in which an individual decides to change her/his identity from masculine to feminine or from feminine to masculine. Doing this entails a permanent biohacking exercise of gender assimilation, a biohacking that consolidates the binarism of sex and gender difference, a biohacking that favors the perpetuation of the colonization of our gendered bodies.

The violence behind it—which Malatino points out as one of the most important conclusions in her paper—is, in my opinion, one of the required points to understand how important is that we may hack and biohack gender to arrive at a posthumanist equality.

It is for this reason that I think that, though Manzocco's book (2019) on transhumanism is one of the most novel and helpful to place the overall debate, it does not convey an issue of great importance such as the post-gender reality with sufficient rigor. His proposal barely addresses the gender issue and, when it does, it does it individually; hence, with no political significance:

In return, we will get greater psychological fluidity and androgyny that we can alter at will; in essence, postgenderism does not demand the cancellation of sexual and gender differences, but rather acknowledgment of the fact that they are the result of choice and not of genetic and cultural imposition (Manzocco 2019: 241).

In other words, there is no ethical involvement or commitment to an ideal to overcome the violence in the construction of gender materiality. The terms "psychological fluidity," "alter at will," and "choice", used by Manzocco, make me contextualize this quote within a psychologizing drift of the world. According to it, experience and knowledge are adapted to the characteristics of psychological management of the dominant model of social organization (De Vos 2013). However, this is the neoliberal capitalism that generates the "humanist" inequality which we are increasingly experiencing today. What is new about a posthumanism or transhumanism based on such premises?

The potentiality of a post-gender world within the framework of a posthumanist vision cannot stay confined in the neoliberal vision whose ideals, concerning gender, are the terms "fluidity" and "choice". This places women and other minoritized subjects within the idea of "aspirational subjects" which we mentioned in the beginning. Moreover, the danger lies in conceiving posthumanism according to the neoliberal rationale which, as Brown points out (2015: 22), leads to:

Both persons and states are construed on the model of the contemporary firm, both persons and states are expected to comport themselves in ways that maximize their capital value in the present and enhance their future value, and both persons and states do so through practices of entrepreneurialism, self-investment, and/or attracting investors.

The subject as part of a conception of "sacrificial citizenship" as Brown calls it. Thus, it is no surprise that, after taking a look at what has been written on biohacking, we mainly find guides that are almost completely dedicated to providing "biohacking strategies to help you achieve exceptional performance". Neoliberalism—as a way to rule everything from the market—has an interest in biohacking—as a disciplinary endeavor based on arranging individuals in a way that turns them into their own bosses—so that every individual produces a free competitive subject who meets the market's interests. In a sexist world with solid sex/gender roles and stereotypes, proposals are normally used to reinforce them.

As Delfanti (2013) and Sánchez Barba (2014: 43) admit, most groups and collectives that could be considered part of the biohacker movement maintain a certain ambiguity with regard to their political stance. Are they a proposal for change to a world that is less dependent on capitalist interests? Or a proposal to better fit in and earn more in that capitalist world? Do they see themselves as a cultural movement rather than a political one?

If continuing with Foucault, the term 'biopolitics' is found in the origins of neoliberalism as the government of life, we will need to find biopolitics and ways to hack the current government of life in order to liberate it. And this is how I understand that biohacking, as a practice of critical biopolitics, may serve a post-sexist posthumanism.

11.4   Feminist Posthumanism and Political Biohacking Ways

To arrive at a post-sexist posthumanism proposal, I am putting forth a concept of governmentality of bodies that transcends the sex/gender binarism that pierces them. The proposal that we elaborate on shall align with a critical conception of biopolitics. The possibilities offered by biohacking practices may be truly promising in this sense. However, this entire constellation of terms—which I present herein for reflection—will only be able to serve a post-sexist posthumanism if it is linked to a collective, non-depoliticized, individualized, and psychologizing action.

My intention has been, in the first place, to make clear that posthumanism has to be a world of consciousness of the human species, and that consciousness is the only thing that can enable us to overcome a world of growing inequality. I understand this inequality is being brought about by a neoliberalism that forces governmentality to follow the market's dictates, as Foucault predicted quite accurately in the 1970s (Brown 2015). The way I see it, a proposal that can consider itself posthumanist must be able to foresee a way to overcome today's problems, the problems of life, and its vulnerabilities (Haraway 2016). For this reason, a posthumanism without a political horizon of transformation of neoliberal capitalism will only be a new stage of evolution of today's humanism.

I have quoted Carbonell, who tells us about a "critical consciousness of species," and I have quoted Haraway, who talks about "rehabilitation" of life as a "making livable again" (Haraway 2016: 33). For both authors, the idea is to create a different world able to reconstruct the conditions of life. For this reason, I find it quite powerful, intellectually speaking, and quite a necessary hook to interpellation when Haraway (2016) poses the following question to us: What kind of care, education, culture, and socialization would you provide to a baby for whom you have to be responsible during the next five generations?

Let's see some of the possibilities.

Osgood et al. (2015) are researching how important it is to take Haraway's reflections and use them to reconfigure the concept and experience of gender in young children (early childhood). In other words, they implicitly understand that we cannot talk about posthumanism without having torn down the walls of gender inequality that have been learned. This requires the creation of a network of new educational practices about social and cultural rules as well as a new conception of who we are as biosocial beings. This reflection on a new education shall include, as Ringrose et al. (2018) also point out, new materialisms and new affect theories (feminist affect theories). This is what Youdell (2017) calls 'biosocial education,' a necessary step to examine the legacy of biosocial research and foster collaborative education among the different disciplines.

As I have already pointed out, the much-needed re-conceptualization of epistemology that Barad, for instance, articulates, must lead us to build bridges of knowledge in a transdisciplinary way. Neuroplasticity, which has been studied for decades and is providing stronger and more promising evidence, should offer new neuro-educational possibilities to the question of going beyond the sex/gender system in education (Reverter-Bañón 2019). Epigenesis is an optimal way to bring new findings that allow us to see ourselves, in this posthuman future, as beings capable of shaping life through collective decisions that we make for ourselves and for the next five (Haraway 2016) or 10,000 generations (Olivares 2014). Both concepts—neuroplasticity and epigenesis—are connected and refer to the permeability of the wall between nature and culture. They also relate to the capacity for performative action which we have mentioned above. The fact that our brain can learn throughout life (neuroplasticity), that environmental influences can affect our gene expression (epigenesis), and that through repeated action we may recode our social identity (performativity), these are important factors to conceive any liberating promise of a posthumanist conception. As Osborne (2015) explains, these scientific findings are essential to understand possibilities when discarding categories that we used to consider permanent:

> Therefore, epigenetics encourages us to view 'natural' traits not as fixed and inherent, but rather as the malleable product of complex interactions between the organism and its environment. As a mechanism through which environment is able to get 'under the skin' to produce complex behavioural traits, epigenetics offers a conceptual bridge that allows Butler's theory of gender performativity to be used productively in conjunction with biological explanations of gender difference. By demonstrating the importance of these types of interactions, epigenetics moves us away from the notion of gender identity as a fixed entity, and towards Butler's argument that gender is instead an unstable constellation of behaviours (Osborne 2015: 508).

However, how can we start doing this? How can we create a beyond-sex-and-gender world? This is the question that is haunting us and which describes quite well the following reflection of Wendy Brown (2005: 98):

> This beyond is a strange place, if it is indeed a place, where it is proposed that the subject and object of the field might be left behind even as the field persists. It is a place where the "what" and the "we" of feminist scholarly work is so undecided or so disseminated that it can no longer bound such work, where the identity that bore women's studies into being has dissolved yet oddly has not dissolved the field itself. Or is it not a place but a time, this beyond of sex and gender? Are we proposing to be after sex and gender, no longer bound by them or perhaps no longer believing in them, and yet, in the peculiar offering that only temporality makes, bringing along what we are after even as we locate it behind us?

Great questions! I understand this moment of dissolution of the sex/gender identity—which has been the basis of feminist criticism, especially since the 1970s—as one of the main cores of the posthumanist proposal that I am suggesting here. Hence, we may call it a feminist posthumanism but not just that, as it shall be inclusive and committed to the diversity of life on the planet.

Some think that this may be done within Modernity and humanism as this same ideal of equality is in fact within the project of Modernity. I am skeptical of this idea. As I mentioned at the beginning of this text, quoting Carbonell (2018), we need a systemic change. Neither Modernity nor humanism, as part of the modern project, are 'ours' any longer in a complete way. They are part of the past. The world and life may be rescued from the conditions of accelerated impoverishment and growing inequalities where we find ourselves. And a big part of these conditions of great vulnerability comes as a consequence of mismanaging the potential of the great ideals of Modernity, as the Frankfurt School has been insisting through all its different generations of thinkers. Rationality, understood in a vulgar dehumanized manner, as Husserl already said in 1936 (1970), ceases to be a useful tool of emancipation and becomes part of the discourse of legitimation of inequalities.

The response of emotions does not help either—to embrace again our lost world to revive it through the family, the people, the nation, the traditions, that which I know because it belongs to me and to us. This response is mostly picked up by the political extreme right, which has, as one of its main pillars, the task to maintain and strengthen this naturalized inequality between women and men.

For all these reasons, I insist on the need to conceive posthumanism as a change of system, episteme, and biopolitical regime. Furthermore, as Haraway (2016) alerts, this change can no longer be a modern 'humanism' since today's issues are planetary, they affect beyond what is human. It's a multi-kind and multi-species question. Today, we are dealing with a world that faces a "double death", as described by Haraway (2016)—it kills life and the conditions that make it possible.

Therefore, the posthumanist proposal that I am presenting has, in the second place, a calling for transformation and emancipation that requires new epistemologies and new knowledge practices. It is no longer useful to separately propose "new knowledges" and "new techniques" that may provide us a future because, for that future to take place, we must structure our present differently. As new chains of affiliation and exploitation are being formed worldwide, we cannot expect a future of solidarity blossoming from this which will put an end to the growing inequalities. We must 'change' the future that this present is preparing for us. Carbonell—an eminent archaeologist who has spent decades studying the Pit of Bones, a 440,000-year-old site in Atapuerca (Burgos, Spain)—tells us that there is no point in understanding our past and present without prospecting our future.

In my reflection, I have borrowed the performativity concept—originally proposed for feminism by Butler (1990, 2015)—as the one to be used for the task of understanding epistemology as a political practice. This point studied and broadened by Barad (2003), helps us propose the term 'performativity' in epistemological and ontological terms. Performativity acts in the subject's education (e.g., through its interpretation within the sex/gender category), but also in the production of bodily matter (linked to the concepts of neuroplasticity and epigenetics). We need to perform the onto-epistemological categorizations in a way that it may serve a posthumanist project where matter, as a biosocial element, maybe reconfigured without creating exclusions, oppressions, or inequalities.

The biohacking practices that we may use to destabilize those onto-epistemological limits that create inequalities shall fit into a horizon of social change and emancipation. For this reason, I link gender biohacking to social and political movements relating to the liberation proclaimed by feminism. This proposal understands that the conditions that shape the bodies are political and that is why using biosocial techniques and actions to institute just and equitable conditions is a political question. As Barad says (2003: 823), "Bodies are not objects with inherent boundaries and properties; they are material-discursive phenomena".

I understand that feminism is a theoretical-practical movement that wants to understand how we have materialized subjectivities and bodies—human and nonhuman—to build a sexist oppressive world towards the subjects and bodies that have been categorized as 'women.' Thus, it is necessary to understand that in order to have a different future, to outline a different way to build life in all its interactions and diversities. To this end, the epistemic system must be changed as well as the biopolitical conditions that govern human relationships and the relationships with all remaining life on the planet Throughout this text, I have attempted to illustrate the impossibility to achieve a post-sexist reality unless we cease to apply the sex/gender identity categories. As some feminisms are claiming (Cooper 2019; Nicholas 2014), the feminist agenda towards equality shall remain limited unless we irradicate the sex/gender identity system. The vulnerability that humanity faces as species requires a consciousness of human species that transcends the onto-epistemic categories that keep us within inequality. And, as I mentioned in the beginning, the sex/gender category is the one that determines our lives in the clearest most unequal way.

How to do this? This is, no doubt, the question that a feminist post-sexist posthumanism should begin to respond to. As a first step, we should no longer apply the sex/gender identification to many of the scenarios that require it today. Much is being discussed on how it may be harmful to health, and it may indeed be so now since the system to conceive and understand bodies is actually under the sex/gender episteme (Pringle 1998). To respond to this, we may say that a post-sexist view must expect a future—which has certainly begun already—where medicine has become more individualized. It is in this example where we may see how scientific knowledge and its biopolitical practices may help shape this future post-sexist world.

In this posthumanist post-sexist world, the present in which we are living today may become the patriarchal prehistory of humanity (Segato 2003). Let's make it so!

References

Asber, Cecilia. 2017. Feminist posthumanities in the anthropocene: Forays into the postnatural. *Journal of Posthuman Studies* 1(2): 185–204.

Barad, Karen. 2003. Posthumanist performativity: Toward an understanding of how matter comes to matter. *Signs: Journal of Women in Culture and Society* 28(3): 801–831.

Braidotti, Rosi. 2013. *The posthuman.* London: Polity Press.

Brown, Wendy. 2005. *Edgework: Critical essays on knowledge and politics.* Princeton University Press.

Brown, Wendy. 2015. *Undoing the demos: Neoliberalism's stealth revolution.* Zone Books.

Bostrom, Nick. 2005. A history of transhumanist thought. *Journal of Evolution and Technology* 14 (1): 1–25.

Butler, Judith. 1990. *Gender trouble: Feminism and the subversion of identity.* New York: Routledge.

Butler, Judith. 2015. *Notes towards a performative theory of assembly.* Harvard University Press.

Carbonell, Eudald. 2018. *Elogio del futuro: Manifiesto por una conciencia crítica de especie.* Barcelona: Arpa Editores.

Carbonell, Eudald, and J. C. Díez Fernández-Lomana. 2019. *Hazte humano: tengas la edad que tengas.* Atapuerca: Diario de los yacimientos de Atapuerca.

Cooper, Davinia. 2019. A very binary drama: The conceptual struggle for gender's future. *feminists@law* 9(1): 1–3.

Delfanti, Alessandro. 2013. *Biohackers: The politics of open science.* Pluto Press.

De Vos, Jan. 2013. *Psychologisation in times of globalisation: Psychologization and the subject of late modernity.* New York: Palgrave Macmillan.

D'Ignazio, Catherine, and Lauren. F. Klein. 2020. *Data feminism.* Boston, MA: MIT Press.

Faubion, J. D. (ed.). 2001. *Michel Foucault, the essential works 1954–1984*, vol. 3. Power London: Allen Lane.

Fausto-Sterling, Ann. 2000. *Sexing the body. Gender politics and the construction of sexuality.* New York, NY: Basic Books.

Ferrando, Francesca. 2013. Posthumanism, transhumanism, antihumanism, metahumanism, and new materialisms. Differences and relations. *An International Journal in Philosophy, Religion, Politics, and the Arts*, 8(2): 26–32.

Fine, Cordelia. 2010. *Delusions of gender: How our minds, society, and neurosexism create difference.* New York: W. W. Norton.

Foucault, Michel. 1969. *L'archéologie du savoir.* Paris: Gallimard.

Foucault, Michel. 2001. *Power: Michel Foucault, the essential works 1954–1984*, vol. 3. London: Allen Lane.

Foucault, Michel. 2009. *Security, territory, population: Lectures at the College de France, 1977–78.* Basingstoke, Hampshire: Palgrave Macmillan.

Gordon, C. 2001. Introduction. In *Power: Michel Foucault, the essential works 1954–1984*, ed. J. D. Faubion, vol. 3, xi–xli. London: Allen Lane.

Grosz, Elizabeth. 1994. *Volatile bodies: Toward a corporeal feminism.* Bloomington: Indiana University Press.

Haraway, Donna. 1985. Manifesto for cyborgs. *Socialist Review* 80: 65–108.

Haraway, Donna. 1991. *Simians, cyborgs, and women: The reinvention of nature.* London: Free Association.

Haraway, Donna. 2006. Interview. Nicholas Gane. "When we have never been human, what is to be done? Interview with Donna Haraway". *Theory, Culture and Society* 23(7–8): 135–158.

Haraway, Donna. 2007. *When species meet.* Minneapolis: University of Minnesota Press.

Haraway, Donna. 2016. *Staying with the trouble: Making kin in the Chthulucene.* Durham: Duke University Press.

Hayles, N. Katherine. 1999. *How we became posthuman: Virtual bodies in cybernetics, literature, and informatics.* Chicago: University of Chicago Press.

Hayles, N. Katherine. 2011. Wrestling with transhumanism. In *H±: Transhumanism and its critics*, ed. G. R. Hansell, and W. Grassie, 215–226. Philadelphia: Metanexus Institute.

Hayward, Eva. 2010. Spider city sex. *Women & Performance: A Journal of Feminist Theory* 20(3): 225–251.

Hester, Helen. 2018. *Xenofeminism.* London: Polity Press.

Hughes, James. 2014. Politics. In *Post- and transhumanism: An introduction*, ed. R. Ranisch and S.L. Sorgner, 133–148. Frankfurt am Main: Peter Lang.

Husserl, Edmund. 1970. *The crisis of European sciences and transcendental phenomenology.* Evanston: Northwestern University Press.

Joel, Daphna, et al. 2015. Sex beyond the genitalia: The human brain mosaic. *Proceedings of the National Academy of Sciences (PNAS)* 112 (50): 15468–15473.

Laboria Cuboniks. 2015. *Xenofeminism: A Politics for Alienation.* https://laboriacuboniks.net/man ifesto/xenofeminism-a-politicsfor-alienation/

Makarychev, Andrey, and Alexander Yatsyk. 2020. *Critical biopolitics of post-soviet.* Lexington Books.

Malatino, Hilary. 2017. Biohacking gender. *Angelaki* 22 (2): 179–190.

Manzocco, Roberto. 2019. *Transhumanism. Engineering the human condition: History, philosophy and current status*. Springer Praxis Books.

Nayar, Pramod K. 2014. *Posthumanism*. London: Polity Press.

Nicholas, Lucy. 2014. *Queer post-gender ethics: The shape of selves to come*. Basingstoke: Palgrave.

Olivares, Lissette. 2014. Hacking the body and posthumanist transbecoming: 10,000 generations later as the *mestizaje* of speculative cyborg feminism and significant otherness. *Nanoethics*, 8(3): 287–297.

Osborne, Jim. 2015. Getting under performance's skin: Epigenetics and gender performativity. *Textual Practice* 29 (3): 499–516.

Osgood, Jane, Red Ruby Scarlet, and Miriam Giugni. 2015. Putting posthumanistic theory to work to reconfigure gender in early childhood: When theory becomes method becomes art. *Global Studies of Childhood* 5 (3): 346–360.

Pringle, Rosemary. 1998. *Sex and medicine: Gender, power and authority in the medical profession*. Cambridge University Press.

Ranisch, Robert, and Stefan Lorenz Sorgner. 2014. Introducing post- and transhumanism. In *Post- and transhumanism: An introduction*, ed. R. Ranisch and S.L. Sorgner, 7–27. Frankfurt am Main: Peter Lang.

Reilly, Colleen. 2018. Gender and bioenhancement. In *Posthumanism. The future of Homo Sapiens*, ed. Michael Bess, and Diana Walsh Pasulka. Macmillan.

Reverter-Bañón, Sonia. 2017a. Performatividad: La teoría especial y la teoría general. *Isegoría*. 56: 61–87.

Reverter-Bañón, Sonia. 2017b. El Neurofeminismo frente a la investigación sobre la diferencia sexual. *Daimon* 6: 95–110.

Reverter-Bañón, Sonia. 2019. The case of gender in moral neuroeducation. In *Moral neuroeducation for a democratic and pluralistic society*, ed. P. Calvo, and J. Gracia-Calandin, 175–192. Springer.

Ringrose, J., K. Warfield, and S. Zarabadi, eds. 2018. *Feminist posthumanisms, new materialism and education*. London: Routledge.

Rippon, Gina. 2019. *The gendered brain. The new neuroscience that shatters the myth of the female brain*. London: Bodley Head.

Sánchez Barba, Gabriela A. 2014. *We are biohackers: Exploring the collective identity of the DIYbio movement*. Master of Science Thesis. Delft University of Technology.

Segato, Rita L. 2016. *La guerra contra las mujeres*. Madrid: Traficantes de sueños.

Stryker, Susan. 2017. *Transgender history: the roots of today's revolution*. New York: Seal Press.

Segato, Rita L. 2003. *Las estructuras elementales de la violencia. Ensayos sobre género entre la antropología, el psicoanálisis y los derechos humanos*. Quilmes: Editorial Universidad Nacional de Quilmes.

Sorgner, Stefan Lorenz. 2014. Pedigrees. In *Post- and transhumanism: An introduction*, ed. R. Ranisch and S.L. Sorgner, 29–48. Frankfurt am Main: Peter Lang.

Spivak, Gayatri Chakravorty. 1993. *Outside in the teaching machine*. New York/London: Routledge.

Stryker, Susan. 2005. Dungeon intimacies: The poetics of transsexual sadomasochism. *Parallax* 14 (1): 36–47.

Tuana, Nancy. 2006. The speculum of ignorance: The women's health movement and epistemologies of ignorance. *Hypatia: A Journal of Feminist Philosophy* 21(3): 1–19.

United National Development Program (UNDP). 2012. Overview of Linkages Between Gender and Climate Change. New York. https://www.undp.org/content/dam/undp/library/gender/Gender%20and%20Environment/PB1_Africa_Overview-Gender-Climate-Change.pdf.

United National Development Program (UNDP). 2020. Tackling Social Norms: A Game-Changer for Gender Inequalities. New York. http://hdr.undp.org/sites/default/files/hd_perspectives_gsni.pdf.

UNWomen. 2020a. Progress of the World's Women 2019–2020a: Families in a Changing World. New York. https://www.unwomen.org/en/digital-library/progress-of-theworlds-women.

UNWomen. 2020b. Women in Politics: 2020b. New York. https://www.unwomen.org/en/digital-lib
rary/publications/2020b/03/women-in-politics-map-2020b.

Vita-More, Natasha. 1983. Transhumanist Manifesto. https://transhumanist.vip/transhumanist-man
ifesto/. (Versions: (1983 v.1, 1998 v.2, 2008 v.3, 2020 v.4)).

Vita-More, Natasha. 2020. *Transhumanist Manifesto*. https://transhumanist.vip/transhumanist-man
ifesto/(Versions: (1983 v.1, 1998 v.2, 2008 v.3, 2020 v.4).

Wittig, Monique. 1992. *The straight mind and other essays*. Harvester Wheatsheaf.

Youdell, Deborah. 2017. Bioscience and the sociology of education: The case for biosocial
education. *British Journal of Sociology of Education* 1–14.

Sonia Reverter is an Associate Professor (Ph.D. in Philosophy) at the Department of Philosophy
and Sociology. Director of Instituto Universitario de Estudios Feministas y de Género (Universitat
Jaume I). Invited scholar from different universities and research centers: Berkeley (USA), University of San Francisco (USA), Gender Institute (London School of Economics), Centre for Civil
Society (London School of Economics), Centro de Investigaciones Filosóficas (Buenos Aires),
Instituto Interdisciplinario de Estudios de Género (Universidad de Buenos Aires). My research
is mainly focused on feminist theory, especially on the new challenges of the feminist agenda.
My goal is to capture the new claims, needs and proposals carried out by feminisms. The intention of this is to articulate a philosophical reflection committed to an emancipating transformation
of the world. In this regard, there is always an idea of a theory as a political commitment in all
my works and publications on cyberfeminism, neurosexism, feminist philosophy, gender violence,
post-gender feminism, transhumanism.

Chapter 12
Ectogenesis and the Ethics of New Reproductive Technologies for Space Exploration

Evie Kendal

Abstract It is often argued that for human settlements to be established off-world we will first need a form of artificial gestation to promote population growth. This chapter considers practical and ethical benefits of embracing ectogenesis technology, both on Earth and for space exploration. It considers the common conflation of artificial gestation with other emerging reproductive technologies, such as cloning and genetic engineering, before considering how ectogenesis might promote the transhumanist agenda, radically destabilising traditional views of family and what makes us human.

12.1 Introduction

By its broadest definition, transhumanism is concerned with the transformation of the human condition and is often focused on applying science and technology to rise above the limitations of human biology (Bostrom 2005). Rather than being at the mercy of "slow, uncontrolled, and unpredictable" human evolution, the transhuman (H+) will be the result of directed evolution driven by human ingenuity and agency (Tirosh-Samuelson and Hurlbut 2016, 7). According to supporters of the transhumanist movement, amelioration of human pain and suffering will be achieved in the future through a variety of means, including "genetic engineering, life-extending biosciences, intelligence intensifiers, smarter interfaces to swifter computers, neural-computer integration, worldwide data networks, virtual reality, intelligent agents, swift electronic communication, artificial intelligence, neuroscience, neural networks, artificial life, off-planet migration and molecular nanotechnology" (More and Vita-More 2003). This chapter will consider the potential contribution of artificial womb technology (ectogenesis) to the transhumanist agenda, focusing on its ability to circumvent the limitations of human biology and fertility. It will also consider two of the specific areas noted above—genetic engineering and off-planet migration—positing that while neither of these are an inevitable extension of

E. Kendal (✉)
Swinburne University of Technology, Melbourne, VIC, Australia
e-mail: ekendal@swin.edu.au

pursuing ectogenesis, both highlight potential benefits of this emerging reproductive biotechnology.

12.2 Ectogenesis and Transhumanism

Ectogenesis refers to the development of a fetus in an artificial womb. This future technology can be separated into two forms: "full ectogenesis", where offspring would be gestated ex utero from conception to viability; and "partial ectogenesis", which could involve transferring an embryo or early fetus from a biological womb into an ectogenesis chamber for continued development. The latter would essentially be an extension of existing humidicrib technology to manage prematurity, while the former would utilise in vitro fertilisation (IVF) and then rely on substantial elongation of the subsequent period of in vitro development currently possible. At present, the longest a human embryo has been kept alive in a petri dish is 13 days, with experimenters in the UK discontinuing the research before running afoul of legislation banning embryo experimentation beyond the 14-day mark (Reardon 2016). Thus, if artificial gestation is to become a reality, it will clearly require both scientific and legal support.

The term "ectogenesis" itself was coined by J. B. S. Haldane in the 1920s who described it as one of the "most important biological discoveries" humanity could make, placing it alongside the domestication of plants, animals, and fungi (Haldane 1923). The concept of growing a human fetus outside the womb has continued to fascinate scientists and science fiction authors ever since. In his 2010 article, "Not of woman born", Eric Steiger lists numerous predictions from scientists regarding when safe and reliable ectogenesis might be realised, ranging from five, to ten, to thirty years in the future (Steiger 2010). Since its publication, the first two of these potential dates have already passed without the advent of this technology. Nevertheless, there have been significant advances made over the last few years, including several successful animal experiments (Partridge et al. 2017; Otway and Ellis 2012), and further steps toward creating bioengineered uteruses (Hellström et al. 2014). Recent scholarship in the area has started to differentiate between different types of artificial gestation, with the use of alternative terms like "ectogestation" and "artificial amnion and placental technology" intended to more accurately convey what many predict the first successful systems will involve (Romanis 2022; Kingma and Finn 2020). However, these predictions typically assume technological assistance for only part of the gestational period, while this chapter is mostly focused on the potential for replacing biological gestation in its entirety. As such, the terms "ectogenesis" and "artificial womb" are retained in this chapter in recognition of the fact many of the benefits outlined here are specific to full ectogenesis. And just as with its predecessor technologies—most notably, IVF and humidicribs – each step toward achieving artificial gestation has been met with controversy and accusations of eugenics and the creation of "designer babies" (Pruchnic 2016). This is where the narrative intersects that of the transhumanist movement.

With its focus on improving the human condition through science and technology, transhumanism provides a useful lens through which to view the potential impact of ectogenesis on future societies and conceptions of what makes us human. While some transhumanists advocate for the abandonment of physical form—for example, through transcending beyond the "encumbrances of the flesh" into virtual existence (Graham 2016, 64)—many others are merely concerned with improving biological systems for optimal wellbeing. As such, the potential contribution of ectogenesis to this goal can be seen to have three distinct beneficiaries: prospective parents, future children, and society. The remainder of this section will focus on the first two of these, including counterarguments against the technology, with later discussion of the third being limited to future off-world settlements. In short, this chapter will consider whether ectogenesis can improve the human experience of procreation for parents, the process of gestation and development for fetuses, and/or the growth of a society in the context of space exploration.

12.2.1 Benefits of Ectogenesis

There are many physiological, psychological, social, and financial benefits that justify the development of ectogenesis, particularly when considering the perspective of those who would otherwise have to biologically gestate offspring (typically prospective mothers or surrogates). These women could avoid the injuries and illnesses associated with pregnancy, as well as the associated loss of income where these conditions impact employment. Ectogenesis might also meet a procreative preference even in the absence of significant health risks or inconvenience, including for women, men, gay couples, and transgender individuals, all of whom may have unique reasons for pursuing this method of reproduction. This is important in the context of establishing a transhumanist society, with diversity of genders and sexualities often considered a desirable condition (see Bostrom's (2003) article "Transhumanist values").

Focusing for the moment on the physical burdens of pregnancy that could be avoided were ectogenesis available, it is noteworthy that many of these are normalised in the context of pregnancy but considered pathological outside of it. For example, nausea, headaches, loss of visual acuity, bone and muscle aches, haemorrhoids, heartburn, etc. all fit within the scope of a "normal" pregnancy, with more serious conditions, such as preeclampsia and gestational diabetes also being relatively common (Zhu and Zhang 2016). Such symptoms can be acutely unpleasant even for those who otherwise value and desire pregnancy. Birth trauma, both physical and psychological, is also a significant risk for both vaginal and caesarean deliveries. Most importantly, pregnant people face a non-trivial risk of permanent disablement and death as a result of their reproductive endeavours. As such, a method of gestating the next generation that could avoid these risks aligns with the transhumanist goal of ameliorating pain and suffering and promoting longevity. Escaping what some have termed the "tyranny of reproduction" thus serves a similar function to the transhumanist ideal of escaping the "burden and curse" of "human biological existence" more broadly (MacKay

2020; Tirosh-Samuelson and Hurlbut 2016). Further, ectogenesis could provide for the needs of prospective parents affected by absolute uterine factor infertility who do not wish to or cannot access uterine transplantation or surrogacy to achieve genetic parenthood (Koplin and Kendal 2020). For cis-gendered men, who have heretofore been dependent on women's bodies to gestate their biological offspring, ectogenesis technology may one day bestow the power to rise above this limitation of human biology—another area of alignment with transhumanist philosophy.

The other potential beneficiaries to be discussed here are the resultant offspring. Embryonic and fetal development are risky undertakings, with genetic abnormalities, toxic in utero exposures, and nutritional deficits all capable of causing significant damage and disability. At present these risks are not equally distributed across demographics, with the latter two tending to cluster among lower socio-economic groups (Kendal 2017). As such, biological gestation may be viewed as a tyranny not only for the "gestators" but also for the gestated, who have no control over the environment that will determine so much of their development. While there are obvious benefits to being able to calibrate the optimal gestation environment from conception, in terms of nutrition, temperature, oxygenation, and avoidance of toxins, the major benefits of ectogenesis, at least in the early stages, are expected to be conferred among infants born prematurely. Alghrani (2018) notes that infants currently born on the "cusp of viability" have high levels of morbidity and mortality, with less than half of those born before 28 weeks' gestational age surviving without disability (111). When coupled with the state's interest in protecting fetal life, this had led some researchers to claim a "moral imperative" exists to pursue research into ectogenesis, especially given its other benefits for prospective parents (Alghrani 2018; Smajdor 2007).

Such techno-mediated reproduction is likely to find supporters among those transhumanists who believe embracing reproductive technologies and rejecting "traditional family values" are relevant indicators of a transhumanist mindset. Bostrom (2005) notes one such figure was early transhumanist, F. M. Esfandiary, (or FM-2030 as he was later known), who espoused the need for transhumanists to support emerging reproductive technologies, including IVF. Regarding the movement's philosophical goal of better understanding humanity's future under conditions of rapidly evolving science and technology, promoting the realisation of a technology with the potential to radically destabilise traditional views of gender and family roles, while embracing the diversity of human experiences of reproduction, fits well with this agenda. What remains to be determined, however, is how the "ectogenetic fetus" will be defined within this new paradigm.

12.2.2   The Ectogenetic Transhuman

Once artificial gestation is perfected the question arises: is the ectogenetic fetus transhuman? Insofar as its gestation has the potential to surpass the limitations of natural biology, the *process* it has undergone surely is, but what about the resultant entity itself? In the absence of any other modifications, the offspring from an artificial womb should be indistinguishable from its biologically gestated peers. However, as Mossman (2012) notes:

> The agenda of transhumanists is to eliminate disease, improve our physical capacities, enhance intelligence, improve our emotional condition, create a happier society for all, and ultimately eliminate aging (247).

Given the justifications for using an artificial womb often include providing an environment free from the risk of infectious disease and promoting fetal survival, there seems to be grounds to consider the ectogenetic fetus a transhuman. That ectogenesis could prevent various disabilities from manifesting, either by preventing harmful exposures or the impacts of maternal dietary deficiencies, also fits some definitions of a transhuman technology. Just as some implants and prostheses are considered to render their users transhuman, including those with defined disabilities, so too might the ability to improve an offspring's "physical capacities" by optimising their prenatal environment. And just as in the former case, such a goal might attract criticism on the grounds of disability discrimination or eugenics (Fletcher 2014). Which leads to the next consideration regarding ectogenesis and transhumanism: the potential not only to avoid prenatal harm and promote survival but to engage in biological modification of human fetuses. This could potentially include those things Mossman notes above, such as enhanced intelligence and avoiding aging.

According to Marchant and López (2012): "While transhumanism involves an often untidy matrix of technologies, processes, goals, and beliefs, the central core of transhumanism is human enhancement" (255). As such, while ectogenesis as a process and the ectogenetic fetus as an entity may both be classified "transhuman", many transhumanists are likely to be more interested in the potential for this technology to mediate human enhancement. This is where ethical challenges become even more complex, in part due to the frequent association of artificial gestation with other—sometimes unrelated—reproductive biotechnologies.

12.2.3   Conflation with Enhancement and Genetic Engineering

Throughout the scientific, bioethical and science-fictional literature focused on ectogenesis there is a common conflation of the potential ethical issues associated with this emerging technology with those surrounding reproductive human cloning and genetic engineering. As these processes typically involve embryonic manipulation, ectogenesis provides no greater opportunity to engage in these practices than already exist using standard IVF. The only meaningful connection is that full ectogenesis would invariably involve the use of IVF, therefore potentially increasing the frequency with which such goals could be readily pursued. Nevertheless, despite not being essential to either cloning or genetic engineering, ectogenesis is often implicated as a slide down the slippery slope of creating a class of genetically modified "haves" who will undermine equity for the unmodified "have nots". Thus, the debate regarding ectogenesis and biological modification of fetuses bears much resemblance to similar arguments regarding the use of transhumanist technologies for human enhancement.

The continual association of cloning and genetic engineering with ectogenesis in the literature and public imagination is problematic, as it perpetuates a belief that pursuing the latter will by default promote the former. Dating back to 1983, a legal article written by Smith II illustrates this belief well in a footnote in which he claims: "If human cloning were to succeed, ectogenesis would have to be perfected to a high degree" (121). At the same time, bioethicist Walters (1982) made a similar claim, suggesting ectogenesis would be necessary "if the cloning of human beings was to go ahead enthusiastically in large numbers" (116). These statements were not true then and are not true now. Mass human reproductive cloning is theoretically possible using existing technologies, it is just very practically difficult given the need to gestate the resultant clones in human wombs. Ectogenesis would merely make it easier to achieve this objective, but it is neither essential nor particularly likely to do so. At least on Earth, the main limiting factor is the predicted cost of the technology should it be made available: it will simply be cheaper to use biological wombs in most cases. The next section will consider how this may not hold in the off-world setting, where issues of scale and practical logistics are unique. As for other potential forms of fetal modification, e.g. surgical interventions, ectogenesis may make such modifications easier to achieve, but it is certainly not a prerequisite (Kendal 2017).

It is important to separate the practical and ethical ramifications of ectogenesis from other reproductive biotechnologies, and likewise the potential for human enhancement that could be facilitated by this technology. The distinction between treatment and enhancement when selecting or altering human embryos for subsequent gestation can be difficult to determine, however, this issue exists apart from whatever gestational method is involved. Nevertheless, the capacity for ectogenesis to ease the process by which a human settlement could be created in space—either in isolation or in conjunction with other transhumanist technologies, such as genetic engineering—warrants further examination. This does not imply these technologies are inextricably linked, but rather that the one may *incidentally* have useful applications for the other, particularly in the space exploration context. Before moving on to this discussion, however, it is necessary to first consider some of the objections to pursuing ectogenesis, with or without other interventions.

12.2.4 Bioconservative Objections

With their focus on preserving "natural" processes, bioconservatists—both secular and religious—tend to oppose emerging reproductive biotechnologies, including ectogenesis. According to Pence (1998), the religious perspective can essentially be summarised as a belief that "humans have no right to change the way humans are created because sexual reproduction was ordained by God", and any attempt to intervene in this natural process would therefore be "sinful" (119). In his article "*Imago*

Dei, DNA, and the transhuman way", Peters (2018) notes some transhumanists use "god-language" to describe the better versions of humanity they envisage, claiming the "perfection H + pursues carries overtones of apotheosis" (357). When accused of "playing God", Peters claims some transhumanists merely "embrace this complaint and take the torch of Prometheus in their hands to storm the gates of heaven" (359). Nevertheless, the objection remains that by fundamentally changing what it is to be human, including at the origins of life, transhumanists risk destroying something sacred. As Bloom (2020) writes:

> The perhaps immediate worry is that transhumanism is equivalent to dehumanization. Put differently, that in its attempts to transcend human limitations it will ultimately lead to destroying of our most sacred human qualities as well as the practical erosions of our social freedoms and free will (7).

The belief that it is immoral or hubristic to replace natural reproduction with techno-mediated reproduction typically coexists with the belief that genetic modification of offspring runs counter to human flourishing. This logic holds that the natural order was created with human flourishing in mind, and as such, any interference must necessarily diminish wellbeing (Burdett and Lorrimar 2019). When it comes to using ectogenesis and/or biological modification to facilitate space exploration and settlement, it is perhaps not too far a leap to say that for the religious bioconservatist, if humanity were meant to settle among the stars they would have been born with the ability to breathe—and breed—there unaided.

For secular bioconservatists, ectogenesis used independently or in concert with other transhuman technologies might serve to undermine core conceptions of what makes us human and promotes solidarity—a diversity that includes imperfections and different abilities (Sandel 2007). For some, the embodiment of pregnancy forms part of this fundamental expression of humanity, both from the perspective of the pregnant person and the fetus. As Parisi (2016) notes, physical touch is "often praised for being the first sense to develop in the womb, grounding the individual's knowledge of the external world in a primary tactile encounter" (88). He further argues:

> In a mediatic ordering of the senses that takes for granted the extension, abstraction, and computerization of vision and hearing, touch has often been framed as having naturally inbuilt bulwarks against such takeovers—what Jacques Derrida termed a "haptocentric intuitionism" that holds touch out as the ultimate and undeceivable guarantor of authenticity … touch often marks the final, irreducible refuge of the human (79).

That an ectogenetic fetus would first experience this touch sensation inside of a machine, rather than a human, could therefore be seen as corrupting this authenticity. The ectogenetic fetus would become the only type of human not to have this shared experience of biological gestation.

When it comes to allowing modifications or enhancements of the ectogenetic fetus, bioconservatists are not the only group concerned about what this might represent in terms of selecting human qualities considered socially desirable, and the risks of imposing an ideal of the "good life" that fails to consider individual and cultural differences (Habermas 2018, 20). When determining which characteristics should be subject to "enhancement", assumptions regarding normalcy and the boundaries of natural limitation might undermine social equality, pathologising attributes that were previously categorised as within the scope of normal species variance, e.g. low normal intelligence, short stature, etc. In a transhumanist future where such traits can simply be eradicated for those able to afford technological interventions, solidarity and acceptance of varying levels of skill among the population could be replaced with intolerance for those unwilling or unable to engage with enhancement technologies. For Fukuyama (2002) this challenge to our "shared human nature" could undermine the foundation of liberal democracy. That humanity could one day be stratified into "enhanced" and "unenhanced" calls into question various commonalities used to promote the idea of equality across all members of the species.

These concerns regarding human enhancement leading to homogenisation of the human species or potential discrimination against the unenhanced are in addition to general safety concerns regarding the application of emerging reproductive biotechnologies. That prospective parents might feel pressured in the future to engage with modification or enhancement of their offspring is also a concern for informed consent in reproductive medicine, whether applied on Earth or off-world. However, there may be higher justifications for pursuing such technologies when considering the unique health risks humans face in space, assuming a future in which off-world migration becomes a real possibility and ectogenesis has been determined to pose no greater risks to fetuses than biological gestation.

12.3   Ectogenesis for Space Exploration

One of the major criticisms levelled at emerging reproductive biotechnologies is that they are unnecessary or frivolous, and therefore detract funding from more "worthy" advancements in health technology. In the case of ectogenesis, such arguments might run: we already have enough/too many people on the planet, we don't need expensive technologies to make more; ectogenesis only caters to the needs of a very small subset of people who are either unable or unwilling to gestate offspring biologically; or, pursuing ectogenesis technology would be a "luxury" focused on creating *new* children, while there are extant children in need of families and other people whose health needs should be given higher priority when allocating finite healthcare resources. However, these common arguments lose traction in the context of off-world migration, where overpopulation is not an issue, all humans would be highly dependent on expensive technology for day to day existence, and society as a whole on population growth for survival.

In the 2005 version of the Transhumanist Declaration, escaping "our confinement to the planet earth" forms part of Article 1, placing it on par with developing artificial intelligence and "overcoming the inevitability of aging" (Bloom 2020). The current version, adopted by the Humanity + Board in 2009, has retained this focus on the need for the human species to leave Earth to reach its potential (Humanity + 2009). Facilitating this process is one of the major contributions ectogenesis could make to the transhumanist agenda, both as an independent technology and one of a suite of potential interventions into human reproduction in space.

12.3.1 Practical Benefits of the Base Technology for Space

Perhaps the most obvious advantage of using ectogenesis to create an off-world human settlement, is the possibility of transporting frozen embryos from Earth instead of large numbers of adult humans. A small crew could be responsible for using the technology to gestate the future settlers and act as guardians until they were old enough to establish the settlement. Although still time consuming, this method would be more efficient than waiting for initial members to create families the old-fashioned way. There is also the benefit that this would free up the initial crew to continue base operations unimpeded by pregnancy and allow the risk of maternal mortality among these limited human settlers to fall to zero. The gender balance among the initial crew would also be less of a concern than if relying on biological wombs to gestate the next generation. From an ethical perspective, removing any perceived or actual duty to procreate among those recruited for off-world migration, avoids a number of potential dilemmas when it comes to reproductive liberty and human rights. While it is likely some settlers will want to biologically gestate their offspring despite any additional risks of doing so in space, the need to ensure this decision is freely made is vital for promoting reproductive autonomy.

In terms of safety, the ability to control gestation under laboratory conditions might be advantageous for fetuses, while avoiding the burdens of pregnancy might have increased significance when considering the strain on the human body of living in space already. Terminating an artificial pregnancy on the grounds of fetal abnormality would also be simpler and safer than terminating a biological pregnancy. In a context where medical resources and expertise are expected to be limited, these advantages could make a significant difference to the feasibility and success of an off-world human settlement.

As noted in the previous section, it is important not to conflate the base technology of ectogenesis with other emerging biotechnologies, including genetic engineering. These technologies can be used independently from each other; ectogenesis need not involve genetic modification, and genetic engineering can occur with biological gestation. But it is equally important to note that while these technologies can be used independently, some practical goals will be easier to realise if they are used in conjunction with each other. For example, combining ectogenesis with genetic modification could help facilitate the creation of "enhanced" humans capable of withstanding higher levels of cosmic radiation than their unenhanced counterparts. It might make especial sense to combine these technologies in the context of space travel—where the benefits of modification are particularly apparent, and some of the usual obstacles to ectogenesis do not apply. These potential benefits will now be discussed in further detail.

12.3.2   Benefits of Ectogenesis for Biological Modification

Konrad Szocik and Tomasz Wójtowicz (2019) claim that the ethical debates regarding human enhancement have failed to engage with the complexities of space travel and the unique threats to health and wellbeing this hostile environment represents for future human settlers. These threats include exposure to radiation and microgravity—factors known to increase life time cancer risk, impact organ function, and reduce bone and muscle density (Blaber et al. 2010; Szocik et al. 2019; Uri and Haven 2005; Mann et al. 2019a)—and to persistently high levels of psychological stress, which has been associated with mental health problems, cognitive impairment, sleep deprivation and immune suppression in astronauts (Marušič et al. 2014; Mann et al. 2019b). At present, many negative health effects of space travel, such as bone demineralisation and muscle atrophy, are handled through strict diets and exercise regimes. Regardless, the prevalence of osteoporosis, orthostatic intolerance, coordination issues and ataxia among returned astronauts remains high (Blaber et al. 2010). Avoiding the additional strain of pregnancy in this situation would be beneficial in its own right, but there is also the possibility artificial gestation could provide targeted physical stimulation and dietary interventions for developing fetuses, at a level too onerous to achieve in the embodied gestational environment. When considering the potentially teratogenic effects of microgravity and occupational exposures on fetal development, it is important to note that ectogenesis provides the only method of separating maternal and fetal exposures.

As noted in the previous section, beyond specialised prenatal diets and physical therapy aimed at preventing developmental disabilities, ectogenesis could be used alongside other reproductive technologies to engage in more extensive biological modification. In the future, it might be possible to use genetic engineering to enhance organ function, metabolic efficiency, and radiation resistance among off-world settlers and their offspring, or influence neurological functions to enhance cognitive ability, psychological resilience, and reduce sleep need (Szocik et al. 2019). These modifications would be of particular value for space exploration, due to the increased physiological and psychological stress of living off-world. In some cases, these modifications would make life easier for settlers, in others, they could be life-saving. The term "modification" is being used here in preference to "enhancement" since, as Smith and Hylkema (2020) note:

> If people simply cannot survive on a world without a given ability, should giving them that ability be considered enhancement in that context? … Settlers on a very different world should not be forced to maintain terrestrial standards of normality, especially when they are not at all practical.
>
> It's worth noting that this is not merely a pragmatic point—it's not just about what we need to do to establish a practical off-world settlement, but about what is in the best interests of the settlers. … no matter how good our technology, the first explorers will face a very high likelihood of death or disfigurement and certainly will suffer from extremely high levels of psychological stress. But this doesn't change the fact that genetic enhancement could significantly reduce their relative risk (224).

When it comes to the welfare of future children in off-world settlements, these authors note arguments in favour of allowing genetic modification might go as far as to suggest that failing to genetically enhance offspring in a context where they will otherwise experience, for example, "an unnecessarily high risk of cancer", is tantamount to child abuse (Smith and Hylkema 2020, 224). Some relevant questions therefore include: if they are to be raised off-world is there a duty to provide the children of settlers whatever supports are available to make this as safe as possible? If so, would these transhumanist interventions be best classified as treatments or enhancements, and how might this distinction impact ethical regulation of these technologies?

12.3.3   The Spacefaring Ectogenetic Transhuman

Any potential boundary between treatment and enhancement blurs in the space environment, where far from creating "superhumans" biological modifications may merely provide some protection against unusually severe risks to health and well-being. As argued in a previous work on the subject, if aimed at mitigating these heightened risks, such measures could be considered prophylactic. When making this determination the following criteria should be considered:

> severity of predicted damage without biological intervention; the nature and acceptability of the proposed intervention; the efficacy of any non-biological alternatives; possible reversibility of the intervention; short and long-term impacts; and incidental risks and benefits beyond the intended purpose (Kendal 2020, 36).

Considering the current mitigation strategies against health threats in space are quite restricted, e.g. protective gear to reduce exposure to radiation can interfere with manoeuvrability and is likely to be particularly burdensome in the context of off-world pregnancy, moves to alter human biology for enhanced radiation resistance and creating alternate sites for gestation appear justified using the above criteria. This is especially the case when considering the interests of future children, for whom adherence to demanding treatment alternatives might be understandably low. Avoiding the need for repeated interventions over the lifespan would need to be ethically balanced against any risks associated with using irreversible interventions, such as various forms of genetic engineering. When considering whether a child born with enhanced capacity to survive in space is unfairly advantaged relative to their non-spacefaring kin, Szocik et al. (2019) claim any "acquired advantages just might barely counteract comparable threats" (73). Modifications that would promote off-world survival can therefore be viewed as compensatory measures intended to treat future generations to live more safely in their environment.

While the above health and safety concerns may justify pursuing the realisation of the spacefaring ectogenetic transhuman, with or without biological modification, there are a number of ethical issues that need to be addressed. These can be broadly separated into safety issues, individual human rights, and species rights. In the first instance, there are ethical concerns developing an artificial womb, whether on Earth or in space, and testing it on a human fetus while it remains an unproven technology. Singer and Wells (2006) capture this dilemma well when they note:

> if it is unethical to attempt ectogenesis in humans until we have a reasonable assurance that it is safe and we can have no reasonable assurance that it is safe until it has been carried out, we seem to be in a classic "catch 22" situation. Work on ectogenesis will remain forever unjustifiable (22).

They claim this vicious cycle might only be broken if ectogenesis is discovered "almost by accident"—in other words, by continuing to refine existing humidicrib technology and attempting to save the lives of fetuses born increasingly prematurely (22). However, the cost–benefit analysis for developing and testing ectogenesis as a means of supporting population growth in an off-world settlement is quite different to that of allowing prospective parents to use it on Earth. Not only are the risks to fetuses in a biological womb expected to be much greater in space than on Earth, but the practical benefits of optimising the gestational environment may be essential for survival. In the space context, ethical concerns regarding the initial testing of ectogenesis may be more relaxed given the risks associated with available alternatives, however, this does not mean space settlements should become the Wild West of embryo experimentation. There may well come a point at which the risks of gestating future children in space, either biologically or artificially, are simply too high to justify the establishment of off-world human settlements at all. Thus, while the threshold for acceptable risks may be higher, it is not unlimited.

Related to the above is the issue of consent. It is clearly necessary that prospective parents be able to give sufficiently informed consent to use ectogenesis to gestate their offspring and understand any other modifications that are being suggested. The right to direct one's own medical treatment is a recognised human right. However, in the case of reproductive medicine, there may be competing interests between prospective parents and future children. Experimental ectogenesis will be quite risky, and it is the resultant offspring who will bear the consequences of any errors in the process. No child ever gives consent to be brought into the world, but there may need to be specific regulations put in place to avoid fetuses being exposed to excessive risk of harm. When it comes to genetic engineering, some bioethicists are also concerned that this might violate the child's "right to an open future" (Davis 2009). The future child cannot agree in advance to the modifications selected by their parents or society, some of which may limit their opportunities in life. In the case of off-world settlers, some modifications might even make it impossible for offspring to return to Earth.

It is interesting to consider whether the spacefaring transhuman might have a plausible right to express their species identity and experience their cultural heritage—in short, whether they have a claim to be able to travel to Earth and engage with "regular" humans. If so, irreversible modifications that would mean they could not survive on Earth or might reasonably be considered a separate species to *Homo sapiens* might be ethically impermissible. There could also be an argument made that humanity should not have to suffer the splitting of its members into two distinct sub-species— essentially H and H+. This would be particularly compelling were it possible to demonstrate a loss of species solidarity and coherence as a result of, for example, genetic engineering for space exploration. All of these issues point to a need to establish ethical guidelines for using emerging reproductive biotechnologies in the context of off-world migration.

12.4 Ethical Regulation of Ectogenesis for Space Exploration

Concerns about transhuman technologies often centre on the risks of promoting eugenics, abandoning human development to be governed by unfettered market forces, and the biopolitical implications of allowing technologies that can erase different abilities (Bloom 2020; Fletcher 2014). However, the common belief that "therapeutic" uses of technology are morally acceptable, while "human enhancement" is ethically dubious, is difficult to defend when biological modifications may confer significant survival advantage, particularly in the hostile space environment. Nevertheless, for those concerned that allowing artificial gestation and/or biological modification of fetuses in the off-world setting will open the floodgates for similar interventions on Earth, it is necessary to develop regulatory systems to prevent gratuitous risk to future generations. As noted in the previous section, rather than focusing on whether the technologies are intended to prevent or cure disease, it might be more meaningful to determine ethical acceptability according to other characteristics, like reversibility, invasiveness, or the efficacy of any alternative health risk mitigation strategies. This might alleviate some of the "slippery slope" concerns bioconservatists have with regards to transhumanism.

Space settlements might provide a useful testing ground for transhumanist technologies, as the scope of influence on human nature is necessarily limited, in part because modifications that are desirable for spacefaring humans may have minimal appeal for humans on Earth. However, Marchant and López (2012) claim that there will always be some people willing to engage with human augmentation and "[a]ll it takes is a relatively small number of initial adopters to try the enhancement technology, and, if it succeeds, interest and support will quickly follow" (261). These authors further claim:

> Any effective restrictions on human enhancement are likely to rely on the legal system to adopt and enforce such prohibitions. While nonlegal measures such as social norms, religious doctrines, or funding restrictions may also slow or hinder the development of human-enhancement technologies, these measures are unlikely to stop such development permanently in the absence of enforceable legal proscriptions (255).

That it would be more challenging to enforce restrictions on biological modification in off-world settlements should be immediately apparent, highlighting why discussions regarding ethically acceptable uses of transhuman technologies are urgently needed before any attempts to establish such settlements occur. The spacefaring ectogenetic transhuman will only come about as a direct result of human intervention and agency—it is definitionally not a natural occurrence. As such, it is relevant to consider what duties and obligations might be owed such an entity, especially considering there is no capacity for an embryo or fetus to give consent for experimentation, gestational or genetic.

This relates to another ethical consideration, the need to establish appropriate consent procedures for prospective parents with regards to the use of emerging technologies that might one day be seen as crossing the species boundary between H and H+. This process would include ensuring the potential consequences of any novel interventions are disclosed, and the ramifications of creating offspring that may differ substantially from typical human children are understood, at least as far as is practicable. Social and legal guardianship issues would also need to be negotiated in advance of any assisted reproductive intervention—for example, if two genetic parents disagree on whether to continue an artificial pregnancy who should decide what happens? Without the typical bodily autonomy arguments in favour of maternal choice, current guidelines may yield little useful counsel here. Justice concerns are also relevant, particularly in a future where not all off-world settlers have equal access to reproductive biotechnologies intended to promote the creation of healthy (space-adapted) offspring. As such, processes for equitable access to high technology reproductive interventions would need to be established from the outset, particularly when such technologies might confer significant survival advantage.

12.5   Conclusion

Assuming a future in which the transhumanist goal of off-world migration becomes a reality, artificial gestation could make a significant contribution to population growth and reproductive safety. Promoting the development of this technology could yield benefits to prospective parents, future children, and societies, both on Earth and in space. However, there are a number of ethical considerations that need to be addressed first, including whether ectogenesis should be allowed to facilitate other interventions into procreation, including biological modification of offspring. Artificial womb technology used independently or in conjunction with other reproductive biotechnologies, can be seen to serve the transhumanist agenda, including through challenging traditional notions of family and community and working to transcend the limits of human biology.

References

Alghrani, A. 2018. *Regulating assisted reproductive technologies: New horizons.* Cambridge: Cambridge University Press.
Blaber, E., H. Marçal, and B.P. Burns. 2010. Bioastronautics: The influence of microgravity on astronaut health. *Astrobiology* 10 (5): 463–473.

Bloom, P. 2020. *Institutions and governance in an AI world: Transhuman relations*. Switzerland: Palgrave Macmillan.

Bostrom, N. 2005. A history of transhumanist thought. *Journal of Evolution and Technology* 14 (1): 1–25.

Bostrom, N. 2003. Transhumanist values. In *Ethical issues for the 21st century*, ed. Adams, F. Virginia: Philosophical Documentation Center Press.

Burdett, M., and V. Lorrimar. 2019. Creatures bound for glory: Biotechnological enhancement and visions of human flourishing. *Studies in Christian Ethics* 32 (2): 241–253.

Davis, D.S. 2009. The parental investment factor and the child's right to an open future. *The Hastings Center Report* 39 (2): 24–27.

Fletcher, D.-J. 2014. Transhuman perfection: The eradication of disability through transhuman technologies. *Humana. Mente Journal of Philosophical Studies* 26: 79–94.

Fukuyama, F. 2002. *Our posthuman future: Consequences of the biotechnology revolution*. New York: Farrar, Straus and Giroux.

Graham, E. 2016. Manifestations of the posthuman in the postsecular imagination. In *Perfecting human futures: Transhuman visions and technological imaginations*, ed. H. Tirosh-Samuelson and J.B. Hurlbut, 51–69. Switzerland: Springer.

Habermas, J. 2018. *The future of human nature*. Cambridge: Polity Press.

Haldane, J.B.S. 1923. *Daedalus, or Science and the future*. London: Kegan Paul Trench Trubner & Co.

Hellström, M., R.R. El-Akouri, C. Sihlbom, B.M. Olsson, J. Lengqvist, H. Bächdahl, et al. 2014. Towards the development of a bioengineered uterus: Comparison of different protocols for rat uterus decellularization. *Acta Biomaterialia* 10 (12): 5034–5042.

Humanity+ (formerly the World Transhumanist Association). 2009. The transhumanist declaration. https://humanityplus.org/philosophy/transhumanist-declaration/. Accessed 1 Oct 2020.

Kendal, E. 2017. The perfect womb: Promoting equality of (fetal) opportunity. *Journal of Bioethical Inquiry* 14: 185–194.

Kendal, E. 2020. Biological modification as prophylaxis: How extreme environments challenge the treatment/enhancement divide. In *Human enhancements for space missions: Lunar, Martian, and future missions to the outer planets*, ed. K. Szocik, 35–45. Switzerland: Springer.

Kingma, E., and S. Finn. 2020. Neonatal incubator or artificial womb? distinguishing ectogestation and ectogenesis using the metaphysics of pregnancy. *Bioethics* 34 (4): 354–363.

Koplin, J.J., and E. Kendal. 2020. Ethical issues in uterine transplantation. *Korean Journal of Transplantation* 34 (2): 78–83.

MacKay, K. 2020. The 'tyranny of reproduction': Could ectogenesis further women's liberation? *Bioethics* 34 (4): 346–353.

Mann, V., A. Sundaresan, and M. Chaganti. 2019a. Cellular changes in the nervous system when exposed to gravitational variation. *Neurology India* 67 (3): 684–691.

Mann, V., A. Sundaresan, S.K. Mehta, B. Crucian, M.F. Doursout, and S. Devakottai. 2019b. Effects of microgravity and other space stressors in immunosuppression and viral reactivation with potential nervous system involvement. *Neurology India* 67 (2): S198-203.

Marchant, G. E., and A. López. 2012. The (in)feasibility of regulating enhancement. In *Building better humans: Refocusing the debate on transhumanism*. ed. H. Tirosh-Samuelson and K. L. Mossman, 255–269. Frankfurt: Peter Lang GmbH.

Marušič, U., R. Meeusen, R. Pišot, and V. Kavcic. 2014. The brain in micro- and hypergravity: The effects of changing gravity on the brain electrocortical activity. *European Journal of Sport Science* 14 (8): 813–822.

More, M., and N. Vita-More. 2003. Principles of extropy, version 3.11.2003. Original version. The extroprian principles. *Extropy* 5(5). Available at www.extropy.org/principles.htm

Mossman, K. L. 2012. In sickness and in health: The (fuzzy) boundary between 'therapy' and 'enhancement'. In *Building better humans: Refocusing the debate on transhumanism*, ed. H. Tirosh-Samuelson and K. L. Frankfurt Mossman, 229–254. Peter Lang GmbH.

Otway, N., and M. Ellis. 2012. Construction and test of an artificial uterus for ex situ development of shark embryos. *Zoo Biology* 31 (2): 197–205.

Parisi, David. 2016. What the surrogate touches: The haptic threshold of transhuman embodiment. *Confero: Essays on Education, Philosophy and Politics* 4: 77–96. https://doi.org/10.3384/confero.2001-4562.161218.

Partridge, E. A., M. G. Davey, M. A. Hornick, P. E. McGovern, A. Y. Mejaddam, and J. D. Vrecenak. et al. 2017. An extra-uterine system to physiologically support the extreme premature lamb. *Nature Communications* 8. https://doi.org/10.1038/ncomms15112

Pence, G. 1998. *Who's afraid of human cloning?* Maryland: Rowman and Littlefield Publishers Inc.

Peters, T. 2018. *Imago Dei,* DNA, and the transhuman way. *Theology and Science* 16 (3): 353–362.

Pruchnic, J. 2016. *Rhetoric and ethics in the cybernetic age: The transhuman condition.* UK: Routledge.

Reardon, S. 2016. Human embryos grown in lab for longer than ever before. *Nature News* 533 (7601): 15–16.

Romanis, E.C. 2022. Assisted gestative technologies. *Journal of Medical Ethics.* https://doi.org/10.1136/medethics-2021-107769.

Sandel, M. 2007. *The case against perfection: Ethics in the age of genetic engineering.* Cambridge: Cambridge University Press.

Singer, P., and D. Wells. 2006. Ectogenesis. In *In ectogenesis: Artificial womb technology and the future of human reproduction,* ed. S. Gelfand and J. Shook, 9–26. New York: Rodopi.

Smajdor, A. 2007. The moral imperative for ectogenesis. *Cambridge Quarterly Healthcare Ethics* 16 (3): 336–345.

Smith, G. P. II. 1983. Intimations of immortality: Clones, cyrons and the law. *U.N.S.W. Law Journal* 6: 119–132.

Smith, Kelly C., and Caleb Hylkema. 2020. "Who's afraid of little green men? Genetic enhancement for off-world settlements". In *Human Enhancement for Space Missions: Lunar, Martian, and future missions to the outer planets,* ed. K. Szocik, 217–27. Switzerland: Springer, 2020.

Steiger, E. 2010. Not of woman born: How ectogenesis will change the way we view viability, birth, and the status of the unborn. *Journal of Law and Health* 23: 143–171.

Szocik, K., and T. Wójtowicz. 2019. Human enhancement in space missions: From moral controversy to technological duty. *Technology in Society* 59: 101–156.

Szocik, K., R. Campa, M. Boone Rappaport, and C. Corbally. 2019. Changing the paradigm on human enhancements: The special case of modifications to counter bone loss for manned Mars missions. *Space Policy* 48: 68–75.

Tirosh-Samuelson, H., and J. B. Hurlbut. 2016. Introduction: Technology, utopianism and eschatology. In *Perfecting human futures: Transhuman visions and technological imaginations,* ed. T. Tirosh-Samuelson and J. B. Hurlbut, 1–28. Switzerland: Springer.

Uri, J.J., and C.P. Haven. 2005. Accomplishments in bioastronautics research aboard International Space Station. *Acta Astronautica* 56: 883–889.

Walters, W.A.W. 1982. Cloning, ectogenesis, and hybrids: Things to come? In *Test-tube babies: A guide to moral questions, present techniques and future possibilities,* ed. W.A.W. Walters and P. Singer, 110–118. Melbourne: Oxford University Press.

Zhu, Y., and C. Zhang. 2016. Prevalence of gestational diabetes and risk of progression to type 2 diabetes: A global perspective. *Current Diabetes Reports* 16 (1): 7.

Evie Kendal is a bioethicist and public health researcher in the School of Health Sciences and Biostatistics, Swinburne University of Technology. Evie's research interests include ethical dilemmas in emerging biotechnologies, space ethics, and public health ethics.

Outro—Short Story: Space Oddity Syndrome by Chris Hables Gray

The Smell of Space

August 23, 2044, 8:03 GMT. Capt. Ortega, Wollstonecraft ship's log, private entry:

E asked me, "How does space feel?" Fucking machine. E can't feel so why does it want to know how I feel? Feel is not for words anyway… but…if I was to use words I'd say, space feels hard and sharp and crisp and I like its smell. It feels cold, but non-judgemental, emotionally. Ultimately, it feels unforgiving. On Earth there is often forgiveness for mistakes. You fall on soft sand and laugh; you startle a rattlesnake and jump back. So the Earthers always think they can fuck up and get away with it. Now Earth is out of forgiveness and they can't adjust.

In space there is no forgiveness. Ever. So we have redundancy, backups. But always there is a disconnect between one's warm body and everything else…. the metal and plastic of the space suit (a tiny space ship), and the spaceship (a giant space suit).

E thinks I hallucinate the smell, insisting it is "a common symptom of space sick-ness." Then it asked me, "How does space smell?".

"Sweet and metallic," I replied. Just like the earliest space walkers said, noticing a difference in the scent of their space suits on returning from the void. "Ethyl formate" the scientists guessed, a common element in cosmic dust clouds according to astronomers. So, deep space dust, not a perfect vacuum after all.

"You expected that smell," E complained.

I insisted the smell is real. But what I should have told E is "Fuck Off!" I also expected euphoria, like Edgar Mitchell, a moon walker from Apollo 14 promised. That great overview feeling, a cosmic Jerusalem Syndrome. Where the fuck is my euphoria?

"Jimmie, come see this."

"Yes, Capitana Maria Encarnación de Ortega."

She pretends to frown as he floats over to look at her monitor.

E. Tumilty and M. Battle-Fisher (eds.), *Transhumanism: Entering an Era of Bodyhacking and Radical Human Modification*, The International Library of Bioethics 100, https://doi.org/10.1007/978-3-031-14328-1

He snorts. "Etiology of Infomania? I can't believe you watch that shit. That's that fake science feed....".

"There's some real science. This is interesting. You know, documentaries. This is about new mental disturbances...disorders?...caused by new techs. So, you know, gamomania, ..."

"Ah, your brother Ángel."

"Yes, this explains part of him...him and games. Or at least labels it. But not just him. They say technology in general always produces new types of anxiety, depression, delusion, mania...."

"Sure, look at us, addicted to space." They can't resist looking at their outside monitors, out at the sparkly black void.

They traverse several hundred kilometers.

"You had a psych profile?" He asks.

"As did you. I've read it." She mimics a classic Scream, palms on cheeks. She can't keep from smiling. "So vanilla," she smirks. "Anyway, my profile mandated that psych program, E."

"And it told you to watch this show?"

"Well, not exactly. When I wanted to talk about my brother, E pushed me to learn more about tech and psych, when its bad... so..."

"Well then, do your homework. Seems like good advice."

"I don't know. It's depressing. The more humans progress the crazier we get. Look at us." She waves around the bridge.

"Fucking bots, the smarter they are the stupider they are."

Several thousand kilometers pass and the silence of space insinuates itself into the command module until, not for the first time, his hand drifts down to her shoulder and the touch becomes a caress. It isn't quiet anymore. Sex between them is new enough that they find the logistics of the free fall love harness challenging. It wasn't just that they weren't yet totally adept at the different options the apparatus afforded, there was a real danger of drifting, indeed hurtling, apart, rendering the necessary physics requirements unmet.

At one pause for adjustments, she whispers, "You know this is just recreational..."

"I prefer medicinal..."

A grunt. "We'll get there."

After they do, they hang, swaying gently, cocooned against the chill.

Extracting herself carefully from their complex tangle, she says, "Wollie, play live Sensanet ™ reporter stream with emo off, volume low."

"Reporting from Lithuania...." The monitor said "This is Cathy Levine from Gedeminas Square. People are organizing a blockade of some sort...Oh, there are reports that... Wait, one of the organizers is speaking...in English it is: "The Russian troops will be here in ten minutes. They are going to try and clear the square. We will confront them without violence, it is our only chance. Form rows around the Ministry. Sit down and link arms, please, fellow citizens."

"Wollie...off. I can't deal with that now," she tells both the spaceship and the sleeping form in the cocoon harness. Then she swims down the hall to the exercise room.

Someone is there, Tim Duncan, part of the cargo, an engineer. The etiquette is unless invited you let the first person use the area. They then limited their workout to 30 min. While many people liked to work out with others, everyone wanted to be alone at times. The 30 min could be precious. But he just nodded when she came in.

"So, any relation...Duncan of the Spurs?"

"No, no relation. I don't even like basketball. You a fan?"

"Yah. A grandfather was from San Antonio."

"Well there you go. I never understood the attraction of sports."

"You sound like Spock, from Star Trek."

"People have called me a Spocker, but I've never even seen the show."

"You really should. It says a lot about the whole drive to go into space."

He looks dubious. "I guess it does influence lots of people. As with the pejorative Spocker...and then the Spockers say it is emotional humans who are mentally ill."

"They sort of have a point. With the emotional humans in command, let's call them the Kirks, we have gone to hell. Look at the world in the last 50 years! Pandemics, the Deluge, war...Tel Aviv!...."

"Okay, stop trying to cheer me up."

She laughs. "Just watch the show, you'll understand us Spacers better."

"You are full of advice," he said, without rancor.

"Probably because I kept getting suggestions from a stupid psych program..."

"E?"

"Yah, but mainly we argue philosophy it seems. And it tells me to research shit. You ever talk with it?"

"Actually, yes, but not to argue philosophy."

"Why then?"

"Well...it started with my space sickness. It ran the biofeedback program, and that really worked for my physical symptoms, so it hasn't become full blown Space Adaption Syndrome. But it turns out there are deeper forms of space sickness... something be-tween dissociative effects ("this isn't real"), loneliness (even with others around), de-pression and/or anxiety...Talking with the program about these things is helpful, for me at least."

"Isn't it some old program? Elmer or something?"

"It is based on a program called Eliza from the last century, but upgraded, obviously. It was written to prove AI psychiatry was nonsensical. Ironic how it turned out. Mainly it does listening therapy with links to information, to cognitive therapy routines (VR games in particular, neurofeedback to baselines on Earth), to other options, including, with a doc's approval, drugs. E isn't judgmental at all."

"Ahhh, now you are talking therapy I can believe in. Do the drugs help?"

"Some. Especially the indica tabs. Mellows me out so I can stand people," he admit-ted.

"So, you love humanity, it is just people you don't like?"

He chuckled. "More or less."

"What about...ah....closeness with people?"

"You mean sex? Is this from my file?" He asked, amused.

"Well, you know I know a lot about you. As Captain I have to make sure...." she shrugged. "And cuddling, hugging, sharing?"

"Okay, Captain. I'll tell you my version, to put you at ease. Physical intimacy is not important to me. I sold it when I was younger and came to realize if I wanted things done right I had to either pay someone or do it myself...and since I'm not into paying strangers to do me, I go my own way. After all I know just what I like."

"I didn't mean to pry..."

"Yes you did. But that's OK. I sometimes find it flattering when people are interested in my psychological states. I wouldn't talk about this if I didn't want to. I don't need cuddles or hugs or someone to make me cum. But I do like talking to smart people, that can't happen alone. That's why I'm talking to you."

"Were you always so...solitary?"

"Probably not, but I was an only child and my parents had issues. I was 9 when my mom died. Oxy. My dad was already on his way, alcohol and anger. People learn this about me and think Black equals poor, that I'm a typical kid who escaped the ghetto. But no, we were solidly middle class until my mom got sick. Then it all went to hell. I ended up on the street, a "rent boy" as we used to say in Hackney. So charming. It wasn't totally my choice, but I was gay already and I sure needed money. Not all the experiences were bad but most were, so it wasn't what I wanted and when I discovered an aptitude for engineering I leveraged that until....here I am."

"Out in the great beyond with the other damaged souls."

He interjects quickly, "Speak for yourself. I'm not damaged. I'm hungry. Hungry for something I could not find on Earth." He looked at her quizzically. "Do you think you're damaged?"

She's quiet for a moment. "No, that was a throwaway line. Sorry. My parents worry that's why I'm out here, addicted to being a hero, and you hear it all the time from Earthers."

"Yes, we do get grief, for coming out here, don't we? And the irony is I loathe being in space. I can't wait until we get to Mars."

"Why, what is the deal with space?"

"It makes my skin crawl. I just try not to think about it. It is so...alien. Space is the extreme environment..."

"And Mars is different?"

"We can deal with Mars. We can remake it. Space is beyond anything we can do to it."

"Hey in Star Trek: Next Generation they discover warp drives are destroying space itself."

"You and Star Trek. I have to say that seems pretty much impossible."

"You see, the warp drive warps space time and sort of wears it out...."

"That's even less convincing."

She laughs, "You're probably right. So, you are into terraforming? No sympathy for the poor dispossessed Martians, if they are there...were there?"

"There is no life there. And even if there was, we're life. We're natural. This is what we do. It will make Mars more complex, more interesting….If there is life there we'll join it. But there probably isn't."

"Well I agree with you. And colonization gives me work. Way more than supplying the Belters."

"Now those people are crazy!"

She nods agreement and then looks at the monitor. "Gotta go send an update to Earth. But Tim, I've enjoyed this. Tomorrow same time? I'll tell you my problems with space."

He shrugs, "I'm always happy to listen to the problems of others. Makes me feel bet-ter."

"See, you aren't supposed to say that."

"Okay," he puts on an unconvincing sympathy face, "Tell me about it."

"I'd like that," with a big fake grin that turns into a genuine smile. "Tomorrow."

The Taste of Space

Wikipedia 2044, accessed August 24, 2044 14:01 GMT

Space Oddity Syndrome: One of a number of psychological disorders specifically brought on by extended space travel. The hard realities of space sometimes lead to great psychological stress, and potentially a psychological break that could end in suicide. It is often exemplified by a tremendous feeling of disconnection, of oddness. Undoubtedly in the same group of disorders as Capgras, Apotemnopiulia, Reduplicative Paramnesia, Ekborm's, Alien Hand, Fregoli Machine, Walking Corpse, Dead Robot Walking, and Robot Corpse syndromes.

> In some ways, the mirror disorder to so-called "Mitchell in Jerusalem Syndrome" after Astronaut Edgar Mitchell's reports of experiencing in space a "profound sense of connectedness…bliss and timelessness…interconnected euphoria" that transformed his life permanently. Termed the "Overview Effect" by Carl Sagan, it can be a healthy perspective—in moderation. But if escalated to mania it is a type of "Jerusalem Syndrome" with dissociative effects ranging from violent mood swings to hallucinations. Related to Stendhal or Florence Syndrome and the recently diagnosed Mars Mania and Lunacy. All defined as variations on hyperculturemia.
>
> Hallucinations are one of the symptoms of space breakdown, and that includes smells. It happens frequently during long stays in Earth orbit, of 37 incidents that have been recorded none yet reported on voyages but there have not been many transits between planets. The assumption is that hallucinations will be a feature of extended space travel as well."
>
> The first hallucinations are often apophenia (perceiving a connection or a significant pattern between random or unrelated things, such as objects, ideas, or sounds). These can also be produced in altered reality games. In space, hearing voices in

the ever-present background noise of ventilation and machinery is typical, but there are also re-ports of seeing a crew mate who isn't there, or in two cases a doppelgänger, as if empty space had begun to fill with mirrors. The theory is that these are generated by a shortage of sensory stimuli.

They meet in the gym again.

"So do you miss Earth," he asks after they exchange nods.

"Well, it is our home…" she sounds unconvinced.

"Not for long." He looks at her. "Earth is so….Old Earth."

They both look out the small window at the little blue ball hanging in space.

They watched the glittery planet for a few minutes, in silence.

"I'm glad you are here," she finally says.

"And why is that?"

"You're so much easier to talk to than E. I can't really connect to it."

"That's probably a good thing. You probably know a major issue in therapy is transference. When the therapist is a digital entity the issue becomes much more complicated. Most common is the reverse of Fregoli Machine Syndrome. Some people call it Eliza Syndrome, actually."

She smiles at that. "No danger of that with me. I strongly prefer people."

He thinks about it, "I'm not sure I do. But I do prefer some people," he glances at her. "And so, if I'm to be your Eliza, why are you here?" He raises his eyebrows.

"Why am I here? In space?" She thinks about it. "I blame our Goldilocks planet. When I was little I loved nature totally. But as I got older I felt the Earth dying. My parents are biologists and in my house there was always this sadness, as things have always been getting worse."

"Things?"

"Yah, funny deflection. Life. The environment. In California it is dying. The fires breaking records every year, the ocean sicker and sicker. The Monarchs gone. It made my parents depressed all the time, drove my brother into games as an escape. He's addicted now. And me…I guess it drove me to space. I always wanted to escape into space, to here, as long as I can remember, but now that I'm here…."

"Now that you are here you are disappointed?"

"Maybe. I like space, I just want to feel more."

"You seem to want the hormones of falling in love, more dopamine and serotonin, not being in love, lower levels of those but more oxytocin."

"You make it sound so romantic."

Duncan shrugged. "Romantic is something I do not understand. I've learned to think about emotions biologically. Maybe you fell in love with space long ago. Now you have a relationship. The hormones change."

She chuckles, but objects, "Having a personal relationship with space sounds a bit whack…What do you call it? God's Eye Delusion? Spaced Out? Cosmic Connection Syndrome?"

He puts on a formal face and says in a monotone, "Such space generated psychological disorders have been postulated. But none have been accepted officially. Commentators note that perhaps such experiences should be seen more as religious or spiritual experiences."

"I'm too pragmatic for religion or spirituality."

Continuing his robot voice, he chants, "Postulate zero-gravity creates new connections in the brain? And the genetic modifications could not change your psychology directly, but your body is different than people without accelerated bone growth and an immune system designed to repair radiation damage. So, your brain and body are changing a great deal being in space. Being dependent on the spaceship for your very existence forces a new intimacy with technology. And the ship has a whole range of sensors beyond yours. You can think of yourself as a new sense for humanity."

"If I didn't know better, I'd say you are being sarcastic."

"That is impossible. Sarcasm is contraindicated for therapy. You must restrain your imagination."

"Sure E. I'll work on that."

Returning to his normal voice, Duncan asks, "So are you happy?"

"Yes. Not ecstatic but happy. I like space. I like the quiet. I like the taste of it. She holds up her hand, "I know, I know…but I do."

"I've heard about smelling space…. I guess since most of taste is actually smell, one could taste it."

"For me, it tastes different than it smells. Less metallic. More electric. Maybe it is just an illusion but I feel a comfort being in space. It is good here on the ship, but even better outside. I don't have the euphoria some people get, or the cosmic religious and political revelations. I just feel comfortable."

Duncan nods. "When I was trying to get over my space phobia, I read this neuro-scientist…Newberg…claimed 'You can tell when you're with someone who has flown in space, it is palpable.' I think for some people you can. You have that air, like a surfer on the beach, of someone who is where they want to be. For the rest of us…."

She smiles. "I can see, most of you colonists are tense all the time. As if you should be ready to throw a patch on a sudden hole in the hull. Won't work, you'd be sucked out before you could move."

"Very, reassuring Captain. Very human of you."

She grins. "Sorta of. You know I'm modified."

"I thought you might be. A number of the other…Martians…have had some gene work done, for rad resistance, bone growth, that Tibetan gene for oxygen…Inheritable."

"Yah, I pretty much got that. And ABC11, low-ordor. You'd be surprised how popular that edit is among Spacers. And a port."

"A port? To jack in?"

"No. Not yet. The option is there but I've not seen the use. Maybe if it was games I was addicted to. My brother uses the date he got his as his "second" birthday. Mine is tiny and actually just slowly vents extra cerebral pressure….turns out hardware works better for that then any drugs."

He nods at her. "Fuck'n spaceborg.

"Fuck'n Martian."

They laugh, strangely pleased.

The Feel of Infinity

October 20, 2044, 7:43 GMT. Capt. Ortega, Wollstonecraft log, personal entry transcript:

"My life on the Wollie has achieved a strange balance. I fuck Jimmie, who is sweet, and good at it. I talk to Duncan, who is logical but somehow empathetic. Or at least he seems so, unlike the psych program, which I've stopped using. Duncan got me to see that my guilt about letting my parents down is complicated. Because they are heavy into saving the planet…Me, going into space, is sort of like deserting, leaving the struggle… But actually they are proud of me. They feel they failed me by not preventing the Deluge. But, of course, it isn't their fault. They spent their lives trying to prevent it. My problem is I lost hope starting in 2020, the first of the pandemics…so many sick, so many dead…nine familia, including both abuelas… They had been raising me while my parents worked and protested. No school for two years. At night I'd go outside, San Juan Bautista is in the country. Beautiful stars. My escape. People say space makes them feel unmoored. That it's brittle. Nothing. Nothing compared to reality. But reality is at its purist in deep space. Overwhelming depth. Literally. It goes on forever. Empty? But it isn't, of course. Outsiders focus on the lack of sound. But we can always hear ourselves. We fill it. We can always hear our own breathing, humming little songs, even our screams if it comes to that….and if the coms are on, others can as well. Yes, it is a hard silence, an aggressive empty silence, a deep cold illuminated by distant cold light. But it is itself, not something humans have made or even effected. It is what it is. Like I said, a reality that won't let you down. It is almost as if…."

"ALARMS/ALARMS/ALARMS"

"Fuck!! Off!"

The cabin exploded with noise.

"This can't help," she screamed over the shrieking alarms. So many systems were in crisis that their various aural signatures "to indicate where the problem was" had turned into an unintelligible 40-tone symphony of disaster.

A third of the sounds went off. "That's the radiation warnings," he shouted back. "I don't know what the fuck the others are."

"Collision…some." Another third of the sounds disappeared. "I don't see any physical shit coming… is it just the…"

"Just the rads gonna kill us?" Inexplicably, he grinned.

Even more surprising, she grinned back. "Rads trigger collision?"

It got noticeably quieter. Raised voices were enough. "Comm failure warnings. Why not just a fucking light?".

"Here's the weirdness. Stability sensors, and others sensors, are reporting problems I just don't feel."

"So any problems we didn't know about?"

"I don't think so Cap. Just the rads coming that…"

"What about the passengers?".

"Shit, they must be freakin' out! I hate sentient cargo."

"No look," she pointed to a monitor on her left. "They're cool. Moving to the shelter."

"Good. It's not that we didn't practice for it. We have contingencies."

"Not for this! This is way more rads than our shelter can take, maybe even the ship. It is a statistically unlikely threat become real."

He shot a quick glance at her and murmured, "I love it when you talk dirty." Louder, he added, "Well Captain, it's a good thing she started out NASA. This babe is waaay over-engineered. Let's work the problem. You want threat?"

"No, you do threat, I'll do options. You reach out to NASA and the other usual suspects and see if they can offer any help."

They looked at each other dubiously. Then they turned to the machines, the comms and monitors. Seventeen minutes passed.

"Threat?" she asked.

"This is a major solar storm. Worst since 1859....the Carington Event, you know...Maybe worse. We got hit by the first photon burst... more photons are coming...the storm is deep in Sol. And heavy particles. We need to shelter in 50 minutes max, but it is worse. We need to harden all the key parts of the ship."

She nodded, unsurprised. "Any help?"

"Comms are spotty. I think we've lost them totally now. But I heard from a number of people. Everyone is very supportive," he said deadpan. "Nice messages from the EU/Japan Space Commission. NASA sent the storm data. Nothing from the big companies ...MuskX, Northrup-DoD, Amazon, PLA-Huawei...No way for them to help, even if there was a profit in it."

"I wasn't expecting any. How could they?"

"What do we got to work with?" he asked.

"There is the shelter. Computers are hardened way above regular specs, thank you NASA. We've got our fuel, water, cargo, passengers. We are locked on this trajectory. Can't run or hide from the main storm."

He nodded this time.

"So, the charged particles are bad for all electrical things...us and the electronics... we've got to defend a massive amount of the ship. Without our electronics we're dead...Options?"

"I came up with only one," she said promptly.

"'k."

"We can use most of our waste water and our liquid hydrogen fuel as an outer shield."

"The fuel? We'll be stuck on Mars... engines cannibalized? How long?"

"I think it's gonna be a long long time..."

He groaned. "Cute."

"We were planning to stay a bit."

"But not with no way back. I'm no NGB! Not doing a Cortez. And you know, with how things on Earth are going, there might not be any ships for 5, even 10 years. We're supposed to go back and then come out again next year. If we do this we're on Mars...maybe forever.

"If we don't we're dead in space. So, what choice do we have?"

There is a long moment of silence. Yes, long…yet just a moment.

"Ok," he conceeds. Just how do we get any of this liquid shit in between us and Sol? It's in pressurized tanks. Two are in the right place, 10%. What we need to survive is running through the recycling system. But the rest…'

"It's all liquid," she remarks. "We can spray it."

"Spray it on the ship? Not much would stick, the outer skin is incredibly resistant to ice forming. You know the…."

"NASA effect. I know. It won't be on our skin, we'll make a cloud. We spray it out and it boils right away, but then immediately freezes…desublimates. Should make a nice cloud of ice particles. They should stay with the ship, stay a cloud, long enough…

"The cloud will have the Wollie's…"'s

"Wollstonecraft. Lady Wollstonecraft"

"…the cloud will have…the Lady's…inertia. And you know she didn't have a title."

"She does now."

"Do we have enough water and still keep enough to live on?"

"It will be close. And smelly."

"You going to tell the cargo?"

She nodded but asked, "The passengers you mean?"

"Why not call them Martians? They like that."

"They haven't even been to Mars yet. Even if they think they're not going back and are going to become transhumans or whatever the fuck they call it this week, I won't call them Martians until they live there. And I'd prefer it if they turned slimy green and grew antenna first!"

He smiled appreciatively and nodded her toward the back of the ship. Then he held up his hand.

"We'll need someone to help us. Someone to connect to the tanks while one of us does the walk and the other does the talk."

She nodded. "I'll get a volunteer."

"How did it go?" he asked on her return.

"All this science, I don't understand…" she sang briefly.

"Come on. Some of them know shit."

"And some don't. But yes, most are pretty sharp and they are all going to follow orders."

"All of them? I thought they were libertarians. Go their own way, get rich."

"I don't think any of them are like old Musk. He is uniquely crazy. And they are all over the place politically. They just agree on Mars. And that on this ship, the Captain's word is law."

He delivered an "Aye Aye, Captain!" with the best snap to attention and salute possible in zero G. "So did you get a volunteer?"

"Yah, Tim Duncan."

"Who's that?"

"You really should learn their names."

"Cargo?"

"The colonists. Anyway, he's the guy you call porn boy….vita-man… You know, the guy who doesn't "do" relationships but loves sex…with himself…" Maria smiles.

"We know way too much about these people."

She scoffed, "You're the one who reads their files like a telenovia just for you."

"As ship's security officer…" he begins.

"You are the only officer. The only crew.

"And we've but one Captain. Who has secured the required volunteer… A surprise, to me, I would have guessed the scientist…. ?

"Pyra? She volunteered but she is too valuable. They have another engineer, younger, female.

"I'm surprised though, he's not your most friendly guy, except to you?" Asked with a tilt of the head.

This is met with a stern look. "Jealousy does not become you."

"Ok. But still he seems a cold fish. Great engineer though… Why did he say he'd do it?"

"He's the one who made the argument that he should go. Because it is logical."

"Well, that sounds better than it's our only hope."

The Stink of Radiation

October 20, 2044 9:01 GMT Maria Ortega, personal emails.

Ms. #1

Familia, we have run into a little trouble up here, I'm sure you've heard. Don't worry. m

Ms. # 2

Carne mio, Ángel mio…don't tell the 'rents but we're in a bit of a jam up here. I'm optimistic we can pull it out but, if we can't, I've no regrets. And, as I'm sharing, I've fallen into a relationship with my mate, Jimmie. He isn't like anyone I've been with. I always knew they were temporary, because none were Spacers. You know I always wanted to be up here. Jimmy is into space, he also is your typical surfer, he's from OB down in SD. Doesn't talk much. 1,000 yard stare sometimes. He fought in Venezuela. One of many things he doesn't talk about. But he is good. Good to fly with and good to me. Makes me laugh. I hope you're getting some laughs bro. besos, m

> "Watch yourself for hypoxia…narrowing vision… colors getting "brittle"… poor judgment and coordination…" he nagged as he helped her into her spacesuit.

"I took the same training Jimmie. You watch my judgement and coordination, and skin going blue…. confusion, cough, rapid breathing and heart rate, shortness of breath, sweating, wheezing…."

He took both her shoulders in his hands and looked at her. "Don't joke Maria. Please. You have 36 min."

"Just don't confuse any spaceflight-associated neuro-ocular effects for hallucinations!"

"Captain and comedian…"

She nodded into his eyes and kissed his nose. She put her helmet on and floated into the air lock.

He shut the door behind her, hit the safety, and then the big red depressurize button. 30 s later the air sucked into the ship, the outer lock opened and she stepped out into the sparkling black.

The first thing she notices is the twinkling. "Looks like dancing fucking fairies," she tells Jimmie cheerfully. "Must be the solar photon storm."

She clumps along the hull with her heavy magnetized boots. When she gets to the outside bay she removes the hose. The nozzle is right there."

"Careful Maria. That connection is not made for space. It'll be brittle from the cold. And the hose…don't kink it…"

"Ok momma," she mutters as she methodically attaches the hose. Then, "Ok. Tell David to start the flow."

She makes sure the nozzle is off until she feels the hose come alive. She looks up a short safety line and turns off her boots so she can float clean. Then, carefully, she opens it slightly. The small fog of liquid transforms into a miniature ice cloud as she is propelled gently to the end of her tether.

"Can you see that Jimmie?"

"I see it Cap. Looks like it is hanging around. This might fucking work."

"Of course it will fucking work," she says, relieved. "I'm the fucking Captain."

"Of course you are."

To herself, she whispers, "Ok, this is still a nontrivial problem. Don't want it to bounce off the ship, don't want to propel oneself into the void, need it at a certain density…." Louder she says, "Let me know how the coverage is."

"Got good visuals Cap."

They work for some time. He gives her small corrections, monitoring the forming cloud. She has to reattach her safety line several times.

Then, "You've to 9 min yet before the storm hits full force."

"Ok, I'm out of liquid anyway…".

"Wait, I can see there's a gap up by the bridge…. Rad monitor confirms… We're getting damage already to the consoles…"

She drifts in silence for thirty seconds.

"I've got liquid."

"It is in the suit. You can't get it out."

"I can a vent the liquid…at the valve."

"Couldn't be enough."

"It is 0.5 kilos water per half hour and whatever I can piss."

"I guess it could be enough… It won't be like spraying a hose…"

"It'll spray out. Vacuum. Just got to get into position."

"Then you'll patch it?"

"No, my fastest patch is 6.7. I'm better off just coming in."

"You'll suffocate. You'll freeze!'

"Nah, I can hold my breath. It'll just be a bit chilly."

"I can't talk you out of this…"

"Tell me exactly where to go."

Once in place, she carefully opened her front pocket and took out the small switch-blade. "I always knew this would come in handy…"

"You and your knife."

She slashes around the waste socket. A small yellow cloud forms. She closes her eyes and voids her bladder."

"Coverage looks good. Get the hell out of there."

One hand pressed ineffectually over the cuts in her suit. She clumps steadily to the airlock. It opens. She staggers in.

He arrives soon after and finds her feebly unlocking her helmet. He takes it off.

"What a show off."

"She smiles weakly. "That's me…"

"Let's go," he says, herding her out of the lock and toward the shelter."

Quietly he says, "We got another gap."

"Okay, give it up."

"There is an opening in the internal shielding, must have slipped a few centimeters, the main net bus is getting fried along a 2 meter length. It fails in 6 minutes."

"Where exactly?"

"Near the engine. We can't get there." He hesitates.

"But David can?"

"He is on his way but…. We've no more shielding. It is all deployed."

"What do we need?"

"80 kilos of water and/or matter. About what his body has…He is hoping to find a different option…"

They enter the tight shelter and she finds a corner monitor and sees the open link.

"David, it's Maria."

"Captain."

"How's it look?"

"Not good. The bus is degraded but it seems to still be working."

"Have you shielded it?"

"Yes, I'm shielding it."

"You're lying on it?"

"I don't see any other options."

She is quiet for a full minute. "I can't think of anything."

"It is what it is."

"Do you want to stay on the comms?"

"No thank you, Maria. I'll do this alone. Thanks for the nice conversations."

"No Tim, thank you."

The link goes dead. Jimmie comes over. "I'll call up his feed."

His vitals spring up on the screen, with one low grade alarm and several regular beeps."

"Turn off the sounds," she asks in a low voice.

The beeps and buzzes cease.

They watch in silence. Heart rate slows, breath rate stops, oxygen in the blood is fine but his temperature climbs relentlessly until heart and respiration flat line. Jimmie turns Tim Duncan's feed off.

His hand finds hers in the silence.

After some time she asks, "You smell that?"

"No. I don't smell anything."

After more silence, she says, "Once we clear the flare someone will have to…"

"I'll deal," he says. "I didn't know him well."

"Thanks Jimmie."

The Look of Death

October 21, 2044, GMT 7:02 Maria Ortega, personal emails.

Ms. # 1.

Familia. I'm ok. Sorry for the cryptic message and the scare. I figured you knew about the flare. We did lose one of the passengers but otherwise no serious injuries. Just some high radiation doses. Space travel is not contraindicated and you know I need my space travel! Anyway, we had to use our fuel to block the radiation so we won't be coming home on schedule. Might take another year or so, you know there aren't that many windows for Earth transit. More later…

Ms. # 2.

Carne mio, hermanito, hang in there. I know you can beat the Game. We are survivors! Abrazo feroz! m.

> They gather in the main bay, the only place the two crew and remaining passengers would fit. On the monitors they watch as the corpse of Tim Duncan, encased in a heavy body bag, is extended out into space and shaken by a robot arm. A cloud of vapor and volatiles escape into space through thin vents and the bag is brought back into the lock.
>
> "What's this about?" Jimmie asks Marie.
>
> "Can't litter in space according to old UN rules, so we couldn't dump his body out there even if he wanted us to. But we can vent the water, so we have less to carry… just his…ashes? Precipitate?"
>
> "Hell if I know. This is my first space funeral."
>
> "Probably not your last." She smiles to take the sting out of it.
>
> "True that, Chica."

"Do you still smell him?".

"Nah, we've got industrial strength air cleaning."

She looks at him dubiously.

The powder safety back in the ship, Marie turns to the colonists, "Pyra, will you say a few words and introduce Tim one last time."

She surveys the people floating in small groups. "We all knew Tim. It is a small ship. But none of us knew Tim. He didn't talk much about himself. Polite we can agree, not warm. So what? He could tell a joke, clean up a mess, sacrifice his life for ours. In the end, what is the sum of us, after all? For many years on Mars he will be remembered. There are now seven dead, sacred offerings to the Red Planet. There will be more. They will be remembered." She held the room in silence, then "Now we'll hear from Tim. His farewell was just updated a few days ago, actually."

All heads swivel to the main monitor as it blinks on, at the bow end. Tim Duncan appears on the screen. He is grayish-black in the harsh artificial light. His whispy beard floats away from his face in the zero gravity. He smiles slightly with his lips, not at all with his eyes.

"I guess it is revealing that I don't find this weird at all. I know…knew?… I'm going to die, sooner rather than later considering I'm immigrating to Mars".

"What do I believe? In energy. Because of it there is something in all the nothing. Emanating from the stars, from our own Sol, in waves and waves of particles full of energy".

"What do I believe? We have the right to terraform Mars. We are a natural process. We have every right to foster life, to spread information. Because that is what life is about, information. Octavia Butler had it right—it is the ultimate value, the only reason to travel to the stars. I believe in spreading information; spreading order. I am an Extropian philosophically, not ontologically. Am I a transhumanist? A posthumanist? I don't know what those mean really. I'm an engineer, not a scientist, but those labels seem profoundly unscientific to me. Can posthumans not breed with humans? Are trans-humans not transexuals? Give me science! Let's wait until we actually are new creatures before we re-name ourselves. We're getting there though, with our gene mods, implants, even the problematical prosthesis of the Belters. Living in space, in the Belt, on Mars we are remaking ourselves. The changes are natural, after all, because we are of nature.

"So no labels for me but I did try to be….tolerant. Forgive this old dead curmudgeon when I was not."

"I have never really liked mass culture, and that includes the Star Trek franchise. So for the longest time I hated being called a "Spocker." But lately I've reconsidered this. On the Wollstoncroft I've had time to watch the shows and Spock is an interesting char-acter and I do relate to him. Logic struggling against human irrationality, as if they were two separate domains. They aren't, of course. Logic is oh-so-human and is not as clear-cut as logical people pretend. Still, I'll own my "Spockness" and say my goodbye:

'Live Long and Prosper!".

"So you pissed up space?" he gave her one of his crooked grins.

"You bet. It was like having an unreasonably long and incredibly skinny penis!"

"How would you know?"

"Fair point."

"Now you've left your mark, you…"

She gave him a quick glance, "You better not say it!"

"beast." He smiled but then looked thoughtful. "Is it like a desecration of space? It has been so unpolluted out here…"

"Unpolluted? Even if you don't count all the photons and cosmic particles and gravitons and normal shit, or that it is expanding at X… The space is full of old TV programs, comet shit, asteroid dandruff… Just spread out. A little human piss just adds to the mix. We can't fuck up space. It is elemental, just like my H_2O and uric acid. It fits right in. I fit right in."

"For now we do. The ship is in surprisingly good shape. We did better than Earth. The flare fried 10% of their electronics, especially Eurasia which took the main hit."

"Great, another disaster. It has been 25 straight years of fucking disaster."

"I know Maria. But it isn't our problem anymore. Getting to Mars, living on Mars, that is going to be years…maybe forever."

"You know we can never go back."

"That's us now, not going back. By necessity."

"Maybe it is a good thing. Trying to make a new world rather than failing to save an old one. We could be happy on Mars."

"Maybe, but I'm not giving up on space." She looked at his blue eyes. "You'll probably love Mars, you're so adaptable. But I won't."

"You don't think so?".

"Mars doesn't call me. Not like it seems to call to the cargo. Space does."

Notes

In 1969 Apollo 10 astronauts reached a top speed for humans of 39,897 km/h (relative to planet Earth). In 2018 NASA's Parker Solar Probe achieved 692,000 km/h, but that was mainly thanks to heliocentric velocity. Speeds on flights to and from Mars will vary, but are expected to surpass 40,000 km/h in many instances. Much depends on the actual trajectory (Earth to Mars travel is ideal only every 26 months but less optimally more often), engine/fuel types, and burn rates.

The physics of liquids in space and in space suits, the dangers and history of solar flares, disposing of bodies from spaceships, and the psychological implications of technology in general and space exploration in particular are based on the sources in the references and the relevant, always well cited, Wikipedia entries. The pioneering work on space sickness and biofeedback by Patricia Cowing first started me thinking about how profoundly our psychology is changed by technology. A big hunk of E's discussion of space psychosis is quoted directly from Kim Stanley Robinson's masterful Red Mars, p. 70, as a homage. Tim Duncan's claim that space is unnatural is borrowed from Bruce Sterling's "Swarm," in Crystal Express, p. 14.

The Wollstonecraft is not unlike Dave Mosher's 2018 proposed "Big Falcon Ship" (https://www.businessinsider.com/spacex-mars-spaceship-big-fal con-rocket-diagram-2018-9) but maybe longer and a bit less chubby. And only the one small diamond window in the gym; small cameras feed various monitors which are always live. The cargo (colonists) are in front and at the very tip, the bridge. The support systems and propellant tanks in the rear, past the central airlock to the outside.

She is named after the incredible Mary Wollstonecraft, author of "A Vindication of the Rights of Women" in 1792. An amazing revolutionary, a cofounder of both modern feminism and anarchism, and mother of Mary Shelley, author of Frankenstein; or, The Modern Prometheus.

While not a transhumanist (nor an extropian from further back), I am in sympathy with some transhumanist perspectives and, unlike many outsiders, I'm well aware of what a complicated and diverse movement it is. It will only grow more so as people actually start living beyond the Earth. My arguments (and Tim Duncan's) for what future humans, and even posthumans, might be called was published in the first issue of the Journal of Posthuman Studies, listed in the Reference section as well. One of the central arguments is that it will probably be extraterrestrial environments that lead to real posthumans: Cyborg sapiens, Marteus sapiens, and the like.

Proposed genetic modifications for space travelers (and potential Martians), include CTNNBI for radiation resistance, LRP5 to accelerate bone growth, ESPA1 which evolved in Tibetans for more efficient oxygen use and ABC11 which is supposed to de-crease "odor production" in the modified.

Hardware modifications seem less useful (as opposed to intimate technologies such as space suits) but brain ports will eventually be developed, especially for gamers and military operators at first. The problem of cerebral pressure damaging eyes among Spacers has already been documented and trepanation ports suggested.

The term cyborg, first coined in 1960 by Manfred Clynes, was explicitly about modifying humans genetically and in other ways, to thrive in space.

This story is second in a series on the psychological implications of postmodern technoscience. The first was published as "The Etiology of Infomania."

Index

© The Editor(s) (if applicable) and The Author(s), under exclusive license
to Springer Nature Switzerland AG 2022
E. Tumilty and M. Battle-Fisher (eds.), *Transhumanism: Entering an Era of Bodyhacking
and Radical Human Modification*, The International Library of Bioethics 100,
https://doi.org/10.1007/978-3-031-14328-1

245